The Convergence of Internet of Things and Cloud for Smart Computing

Intelligent Signal Processing and Data Analysis

SERIES EDITOR

Nilanjan Dey, Department of Information Technology, Techno India College of Technology, Kolkata, India

Proposals for the series should be sent directly to one of the series editors above, or submitted to:

Chapman & Hall/CRC
Taylor and Francis Group
3 Park Square, Milton Park
Abingdon, OX14 4RN, UK

For more information about this series, please visit: www.routledge.com/Intelligent-Signal-Processing-and-Data-Analysis/book-series/INSPDA

The Convergence of Internet of Things and Cloud for Smart Computing

Parikshit N. Mahalle, Nancy Ambritta P.,
Gitanjali Rahul Shinde and Arvind Vinayak Deshpande

CRC Press
Taylor & Francis Group
Boca Raton London New York

First edition published 2022
by CRC Press
6000 Broken Sound Parkway NW, Suite 300, Boca Raton, FL 33487-2742

and by CRC Press
2 Park Square, Milton Park, Abingdon, Oxon OX14 4RN

Library of Congress Cataloging-in-Publication Data
A catalog record has been requested for this book

ISBN: 978-1-032-03804-9 (hbk)
ISBN: 978-1-032-03805-6 (pbk)
ISBN: 978-1-003-18909-1 (ebk)

Typeset in Times
by Newgen Publishing UK

CONTENTS

AUTHORS

Parikshit N. Mahalle obtained his B.E. degree in computer science and engineering from Sant Gadge Baba Amravati University, Amravati, India and M.E. degree in computer engineering from Savitribai Phule Pune University, Pune, India. He completed his Ph.D. in computer science and engineering with a specialization in wireless communication from Aalborg University, Aalborg, Denmark. He was a post-doctoral researcher at CMI, Aalborg University, Copenhagen, Denmark. Currently, he is working as professor and head in the Department of Computer Engineering at Sinhgad Technical Education Society's (STES's) Smt. Kashibai Navale College of Engineering, Pune, India. He has more than 20 years of teaching and research experience. He is serving as a subject expert in computer engineering, Research and Recognition Committee at several universities like SPPU (Pune) and SGBU (Amravati).

He is a senior member of IEEE, ACM member, life member of both CSI and ISTE. Also, he is a member in the editorial board of journals, *IEEE Transaction on Information Forensics and Security* and *IEEE Internet of Things Journal*. He is a reviewer for IGI Global – *International Journal of Rough Sets and Data Analysis* (IJRSDA), Associate Editor for IGI Global – *International Journal of Synthetic Emotions* (IJSE), and *Interscience International Journal of Grid and Utility Computing* (IJGUC). He is a member of the Editorial Review Board for IGI Global – *International Journal of Ambient Computing and Intelligence* (IJACI). He is also working as an associate editor for IGI Global – *International Journal of Synthetic Emotions* (IJSE). He has also been a technical program committee member for international conferences and symposium like IEEE ICC, IEEE INDICON, IEEE GCWSN, and IEEE ICCUBEA.

He is a reviewer for the Springer journal *Wireless Personal Communications*, reviewer for Elsevier journal *Applied Computing and Informatics*, and member of the Editorial Review Board of the journals IGI Global – *International Journal of Ambient Computing and Intelligence* (IJACI) and *Global Research in Computer Science*.

He has published more than 150 research publications having 1,372 citations and H index 16. He has been an author/editor in five edited books published by Springer and CRC Press. He has seven patents to his credit. He has also delivered an invited talk on "Identity Management in IoT" to Symantec Research Lab, Mountain View, California. He has delivered more than 100 lectures at national and international level on IoT, Big Data, and digitization. He has authored 11 books on various subjects. Some of them are *Context-Aware Pervasive Systems and Application* (Springer Nature Press), *Design and Analysis of Algorithms* (Cambridge University), *Identity Management for the Internet of Things* (River Publications), *Data Structure and Algorithms* (Cengage Publications), and *Programming Using Python* (Tech-Neo Publications MSBTE).

He was the Chairman of Board of Studies (Information Technology), SPPU, Pune. He is now the Member – Board of Studies (Computer Engineering), SPPU, Pune. He has been a member of the Board of Studies at several institutions like VIT (Pune), Govt. College (Karad), Sandeep University (Nashik), Vishwakarma University (Pune), and Dr. D. Y. Patil International University (Pune). He has also been a technical program committee member for many international conferences.

He is a recognized Ph.D. guide of SSPU, Pune, guiding 7 Ph.D. students in the area of IoT and machine learning. Recently, two students have successfully defended their Ph.D. He is also the recipient of "Best Faculty Award" by Sinhgad Institutes and Cognizant Technologies Solutions. His recent research interests include algorithms, Internet of Things, identity management, and security. He has visited a few countries like Denmark, France, Sweden, Germany, Austria, Norway, China, Switzerland, and Singapore.

Nancy Ambritta P. graduated in computer science and engineering from Anna University, Tamil Nadu, India, in 2010 and has her master's in computer engineering from Smt. Kashibai Navale College of Engineering, Pune, India, in 2015. Her research interests are Cloud security and the Future Internet. She has about three years of experience in the software engineering field and 3.5 years of experience in teaching and research. She has published research articles in international journals and conferences. She is a reviewer for the Springer journal *Wireless Personal Communications*.

Gitanjali Rahul Shinde has overall 11 years of experience, presently working as SPPU-approved Assistant Professor in Department of Computer Engineering, Smt. Kashibai Navale College of Engineering, Pune. She has done a Ph.D. in wireless communication from CMI, Aalborg University, Copenhagen, Denmark on "Cluster Framework for Internet of People, Things and Services" and awarded the Ph.D. on May 8, 2018. She had obtained M.E. (computer engineering) degree from the University of Pune, Pune, in 2012 and B.E. (computer engineering) from the University of Pune, Pune, in 2006. She has received research funding for project "Lightweight Group Authentication for IoT" carried out in SPPU, Pune. She has presented a research article in the World Wireless Research Forum (WWRF) meeting at Beijing. She has published more than 40 papers in national and international conferences and journals. She is author of three books and also editor of book *The Internet of Everything: Advances, Challenges and Application* (De Gruyter Press).

Arvind Vinayak Deshpande has obtained B.E. in electrical engineering and Ph.D. in computer engineering from Shivaji University, India. He is now the Principal of Smt. Kashibai Navale College of Engineering since December 2003. He was selected at ISVOR-FIAT International Italy as a Master-Trainer in 1998. He was selected as chairperson to conduct a session on Image Compression in the international conference held at Marbella, Spain, in September 2004 by IASTED, Canada. He was a member of the Board of Studies (BOS) of Computer Engineering, University of Pune, India. He has signed a memorandum of understanding (MOU) with a Rwanda nonprofit organization (NGO) for setting up an engineering college and MBA institute at Kigali in Rwanda (Central Africa). He is chief co-coordinator of the Indian-French student organization. He has received several awards for his contribution to engineering education at the state and national level, which has been recognized by Indo-American Society, Indian Society for Technical Education (ISTE), IETE, and other voluntary organizations. He is a recognized Ph.D. guide in computer engineering at SPPU.

PREFACE

Failure will never overtake me when my definition to success is strong enough!

Dr. A.P.J. Abdul Kalam

The IoT–Cloud convergence has occupied the limelight in recent years due to the emerging demands in IoT applications. The IoT applications have evolved from merely simple standalone applications such as home automation, process automation, or care for elders to advanced ones that include oil and natural gas rig safety, IoT for agriculture, and many more. The resource requirements for processing standalone applications can be sufficiently met by the participating devices themselves. However, advanced IoT applications have increased demand for resources due to huge volumes of data coming in from various sources and the need for huge computations. This cannot be met solely by the participating devices, thereby leading to the convergence of Cloud into the applications.

The use of converged technology is growing enormously as there is an enhancement in smart devices, communication, Internet of Things (IoT), and Cloud computing. The convergence of IoT and Cloud computing is required for various purposes in real-time scenarios. The fundamental requirement of IoT is resource sharing. Cloud computing not only works upon the principle of sharing resources but also emphasizes maximizing the availability of resources. One of the key features of Cloud is the virtualization of underlying physical devices, which makes it easier for users to share the same hardware in a seamless manner. IoT has many mobile users and devices. Services should be made available to these devices/users independent of the location. Cloud being location independent allows services to be accessed from anywhere and from any device over the internet. The mentioned advantages assure that the convergence of IoT and Cloud can lead to many successful developments in the field. Considering the above-mentioned information, it is rightful to provide interested readers with a book that efficiently presents the various technologies and methods in an organized manner, enabling users to move gradually from the basic to advanced concepts. This will enable novice readers and researchers find all necessary information related to IoT application development, which can lead to many productive developments and findings in the area.

This book is a one-stop shop that offers the readers everything they need to know or use in a real-time IoT application development activity. The book offers a basic understanding of the IoT architecture, use cases, smart computing, and the associated challenges in designing and developing the IoT system. This provides a strong foundation to a novice reader. All the technical details that include information about the protocol stack, technologies, and platforms used for the implementation are explained in a clear and organized format. Various aspects pertaining to the IoT–Cloud computing platform along with an introduction to the offloading and resource sharing mechanisms are discussed in the book, which helps the reader to gain an in-depth understanding about the internal working of the applications/systems. Techniques and case studies that include smart computing on the IoT Cloud models along with test beds for experimentation purposes are introduced to the reader. An outlook from where the readers can build upon and work toward developing their applications is presented in the book.

This book provides challenges involved in IoT application and discusses various technologies, protocols, and platforms that are used in IoT application development. This book also summarizes various testbeds that is a part of the FIESTA-IoT project that aids in experimenting new approaches toward the development and deployment of IoT applications. A few key features of this book are as follows:

- Discusses the broad background of IoT and its fundamentals
- Discusses the broad background of IoT–Cloud convergence architectures and its fundamentals along with resource provisioning mechanisms

- Emphasizes the use of context in developing context aware IoT solutions
- Discusses various technologies, methodologies, and approaches that play a prominent role in smart computing on the IoT Cloud architecture
- Presents case studies of a few application areas of smart computing over IoT Cloud
- Presents a novel C-model that explains the IoT application development phases in a simple yet effective way
- Presents a simplified convergence model that depicts the role of Cloud in an IoT application.

In a nutshell, this book presents all information (basic and advanced) that a novice and advanced reader needs to know regarding IoT application development. The book also recommends the use of appropriate technology for better development of applications. The book also contributes to social responsibilities by laying down the foundation for the development of applications that can help in making day-to-day activities easier by meeting the requirements of various sectors such as education, healthcare, agriculture, and other vital aspects of human lives.

ACKNOWLEDGMENTS

We would like to thank many people who encouraged and helped us in various ways throughout this book, namely our colleagues, friends, and students. Special thanks to our family for their support and care. We are thankful to the founder president of STES, Prof. M. N. Navale, founder secretary of STES; Dr. (Mrs.) S. M. Navale, Vice President (HR); Mr. Rohit M. Navale, Vice President (Admin); Ms. Rachana M. Navale, our Principal; Dr. A. V. Deshpande, Vice Principal; Dr. K. R. Borole; and Dr. K. N. Honwadkar for their constant encouragement and great support.

We would like to thank the Prof. Dr. Nilanjan Dey for his continuous support and constructive guidance. We are also very much thankful to all our department colleagues at SKNCOE, for their continued support, help, and keeping us smiling all the time.

Last but not the least, our acknowledgments would remain incomplete if we do not thank the team of CRC Press, Taylor & Francis Group, who supported us throughout the development of this book. It has been a pleasure to work with the team and we extend our special thanks to the entire team involved in the publication of this book.

Parikshit N. Mahalle
Nancy Ambritta P.
Gitanjali Rahul Shinde
Arvind Vinayak Deshpande

INTRODUCTION

1.1 OVERVIEW OF IOT

The term Internet of Things (IoT) was first introduced in 1999 by Kevin Ashton in the *RFID Journal*. Later, the International Telecommunication Union (ITU) came up with a unique idea for connectivity among the devices: Anyone will be able to have connectivity with anything (device to device or human to device) at any time or any place (with any other device or device in commute). In the past, when this idea was discussed in the industry, the whole idea did not seem realistic to them. However ITU and small business ventures incorporated some sensors into the devices to make everyone see the evolution of the internet along with the combination of technology and vision. This new vision has the ability to change our day-to-day life forever. This is one of the biggest reasons why IoT products and their associated services will become popular in the society. IoT will greatly impact every part of the life – work, home, and social [1].

The definition of IoT is perceived differently, as many researchers have different ideas about IoT. Because of the different definitions, it has become truly difficult to comprehend the exact definition of IoT. On the same lines, many scientists, researchers, and standardization bodies are working on it but from their own perspective. This is creating more perspectives in addition to what is present right now. A combination of these merged perspectives from the abovementioned bodies can help us understand IoT and its ecosystem. IoT can be partitioned into three different dimensions on the basis of multiple visions:

1. The first dimension consists of things which include key chains, portable medical devices, or watches.
2. The second dimension is internet or connectivity-oriented dimension. This includes communication and connectivity among things.

3. The third dimension is semantic. It means data collected by various devices should be provided with some sort of technology in order to store, collect, sort and interpret that data.

The things can be further partitioned into different categories. Day-to-day objects like books, wallets, and carts can be connected to sensors, such as a temperature sensor and motion sensor. The others can be combined with appliances like a refrigerator or air conditioner. Sensors have been in use for a long time, but its connection with IoT can be helpful. When the term IoT is referred, it means connectivity among devices without human intervention. A vast amount of literature is available on IoT, but the difference in their perspectives causes confusion.

1.1.1 Technical Building Blocks

In the following section, the taxonomy of IoT components is presented. There are three main components of IoT [2,3]:

1. Hardware, which consists of sensors and actuators
2. Middleware, which is provided for the purpose of data analysis
3. Presentation, which is visualization of information made easy with the help of this layer.

Hardware, that is sensors and actuators, are the main components of IoT. Sensors are used to collect information on parameters like temperature, pressure, or humidity. Actuators are programmed to take action based on the value sensed by the sensors. These sensors can communicate through various wireless communication technologies like radio frequency identification (RFID), WiFi, and Bluetooth. These sensors need to have unique identification to be deployed in IoT application. WSN is the basic network in IoT. Data sensed by these sensors are stored and processed using Cloud computing. A few technical building blocks of IoT are discussed as follows:

1.1.1.1 Radio Frequency Identification (RFID)

RFID is one of the technologies which can be embedded into the microchip and can be used in wireless communication. Identification of the things that are attached to it is a major task. There are two types of RFIDs:

1. Active RFID: This RFID is powered. It means that it has its own power source. It can also communicate if necessary. Monitoring of valuable cargos is achieved by the active RFID tags.
2. Passive RFID: This RFID does not use battery. It makes use of readers' signal to identify the ID for the RFID reader. The application of this RFID can be seen in retail, transportation, access control etc. These tags are used in various applications like road toll, banking cards etc.

1.1.1.2 Wireless Sensor Networks (WSN)

WSN consists of a number of intelligent sensors. These sensors continuously gather information after which processing is done on the collected data. The pre-processed data is used for analysis so that valuable information can be extracted from the collected data. Active RFID and lower end WSN nodes serve the same purpose in terms of processing capability and storage. The challenges that come with WSN are multidisciplinary in nature and so is the potential of this kind of network. Sensor data that is gathered from sensors is sent to the centralized or distributed systems for analysis. Recent advancements in technology have made this wireless communication very easy.

1.1.1.3 Addressing Schemes

Any IoT system should be able to uniquely identify things. This step is very crucial in making that IoT system a success. This will help in identification of billions of devices uniquely. Also, the management of these devices with the help of internet will be possible. A major critical aspect of creating a unique address is as follows:

- Uniqueness of address
- Reliability
- Persistence
- Scalability

Every "thing" that is connected in the system and those that are going to join should have a unique address, location, identity, and functionality. Currently the IPV4 addressing scheme is being used, but as the number of the devices increases, it becomes impossible for IPV4 to give unique ID's to each and every device. The solution to this problem is the IPV6 addressing scheme. However, this will only solve the problem to some extent. The heterogeneity of the devices, their associated data types, and the operations on them still remains a major concern. In order to channel the data traffic ubiquitously and relentlessly, persistent network functioning can be used. Even if the TCP/IP is taking care of the efficient delivery from source to destination, IoT is still facing a bottleneck at the interface of the gateway and the sensors. Scalability of the available address should be sustainable. If at all another network is required to join in, then the performance of the existing network should not be hampered in any way. To solve these problems, Uniform Resource Name (URN) system was developed, which is being considered as the base of the development in IoT. This system creates replicas or mirror images of the devices that is more easily accessible through the URL. As large volumes of data are being created, the metadata can help in the transfer of this data. IPV6 is also a good option to be able to uniquely identify things and monitor and control them. Another important task is the development of lightweight IPV6 for home appliances.

1.1.1.4 Data Storage and Analytics

The unprecedented outcome of any emerging field is the generation of a large amount of data. The storage and management of the data, ownership, and expired data issues have gained importance. According to statistics 5 percent of the totally generated energy is being used by the internet. As days go by, this statistics will change; it simply means the demand will increase. Hence the harvested energy using data centers needs to be figured out plans for conserving energy. The storage and use of data should be done smartly. It is imminent to design AI algorithms that could be implemented in a centralized or a distributed way. Fusion of algorithms needs to be designed to interpret the collected data. In the state-of-the-art various methods like nonlinear, genetic algorithms, temporal machine learning methods based on evolutionary algorithms, neural networks, and other artificial intelligence techniques are required to gain control of automated decision making. These systems have abilities like interoperability, integration, and adaptive communications. They also have modular architecture in hardware as well as in software parts, which is required for any IoT application.

1.1.1.5 Cloud Processing

The data is useless if there is no provision to store, collect, or analyze that data. This is made possible with the help of Cloud computing. The collected data needs to be analyzed to deduce if any action is necessary, depending on the results. For example, when a person is reaching home from work, then the thermostat can be turned on depending on the distance between his current location and home, so that when that person reaches home the temperature will be perfect. On reaching the front gate, the gates will be opened and the person can drive in. This is made possible because things collaborate and take actions if necessary, which is made possible due to Cloud processing. The thermostat and the front gate have sensors which sense the current situation and act on it with the help of Cloud.

1.1.1.6 Security

Security is very important in any domain. In IoT the sensors, their connecting interfaces, communication among them, infrastructure for the system, Cloud processing, and storage require security at every step. When IoT products were being developed, security was not the biggest concern at the time. Because of this flaw many companies still haven't implemented the IoT measures. As time progressed, it became evident that security in the system needs to be strengthened. As hackers are tampering with vehicle control systems, for example, more stringent security measures are needed to be implemented. Hence many companies are asking for IoT products that have strong security systems that will be difficult to breach.

1.1.1.7 Visualization

Visualization is a very important milestone in the creation of an IoT application. This visualization helps users to easily communicate with the environment. The displays (phones, tablets) nowadays are designed to be more intuitive so that users will be able to reap full benefit from it. For a layman, it is very important to get the information needed in a more precise, compact, and easily understandable format. That is why visualization is very critical. As displays are evolving from 2D to 3D versions, more information can be made available to the user. This will in turn help policymakers to transform data into meaningful knowledge. This consists of visualization and detection of the associated raw data.

1.1.2 IoT in Business

IoT can change the perspective of business approach by providing IoT connectivity. IoT will help businesses track and effectively monitor assets in real time with the help of sensors [4]. Various industries are shifting from manual working toward IoT due to automation and smart applications.

1.1.2.1 Intelligent Transportation

Transportation is one of the important domains where IoT can make a difference. Using sensors, information about vehicles such as their location, route, their physical condition, and many more can be retrieved faster and without human intervention. This information can be of help for safe and smart transportation. Further IoT can be used in car networks to share high-rate multimedia data. This type of network is called as vehicular ad hoc networks (VANETs). Device-to-device communication (D2D) is the pillar of such networks. In D2D communication, devices will be able to communicate with very less involvement of the network. This solves the latency issues to some extent where vehicles will be able to communicate with each other directly using sensors.

1.1.2.2 Smart Homes/Buildings and Monitoring

Sensors come in handy for the prevention of hazardous situations. The sensors can be installed at home to monitor the situation. For example, environmental sensors can be used to check the quality of air in the house such as levels of carbon monoxide concentration, colour, gas leaks, humidity, hydrogen sulfide levels, and temperature remotely. Home automation is the best example of IoT application in the households. With the help of IoT lights around the house can be controlled. Also, if there are elderly people at home then incorporation of IoT in day-to-day life can make life better. Continuous monitoring of elderly people will become very efficient with IoT.

To provide full safety to buildings, IoT can be incorporated in the buildings. Some of the startups are developing sensors that can be embedded into the foundation of the building to monitor the consistency of load bearing, heating systems, and lift.

1.1.2.3 Smart Grids

In early days when the smart grids were installed, the main task of the grid was to control supply and if necessary cutoff the supply of power as per instructions. But incorporating the IoT in the grid can provide the grid with the ability to execute a wide array of functions. These grids will be able to remotely monitor not only the power requirements but also the gas or water requirements. In future smart grids will be able to replace sensors, which will be able to reduce cost by relaying voltage and current measurements.

1.1.2.4 Environment Observation and Forecasting

Growing urbanization and harmful human activities are proving to be detrimental to nature. To preserve the environment and all the life it holds, IoT can be very helpful. There are some species that are on the verge of extinction. Tracking of such species can be done with global positioning system (GPS)-enabled sensors. Preserving such species will be a little bit easy with the sensors. Long-range radio transmitters can be used to monitor the ground habitats of the species. Because of the extinction of some of the species, they have been termed as exotic. Trespassing has also increased on the habitats for the possession of such animals. IoT can help to ensure the safety of such animals.

Deforestation is also a grave threat to animal life. Because of wildfires or human mistakes, it is rapidly increasing. IoT sensors can be deployed in the dense forests, which can alert the system just in time if any natural tragedy is about to occur. IoT can also help predict earthquakes or avalanche, so that the population at risk can be removed from the place as soon as possible.

1.1.2.5 Smart Agriculture and Farming

Water is one of the most important needs of life. It has to be conserved because reservoirs are depleting at faster rate. According to statistics 70 percent of the fresh water is required for farming purposes, but 60 percent of that goes to waste because of the failure in irrigation system and faulty agricultural techniques. Sensors can be installed in the irrigation system to detect the failures in the system. In this way, whenever there is a failure in the irrigation system, immediate action can be taken to stop the water wastage. IoT systems can be developed for vineyards to sense temperature, soil moisture, and solar radiation and alerts can be sent to the farmers in case of a dangerous situation.

1.1.2.6 Health Care

There are numerous fields available in which IoT can be applied. IoT can lend a helping hand in the revolutionization of the health care industry. If a patient needs continuous monitoring then the common practice in health care industry is that a physician is required to continuously monitor the patient. But recent trends in WSN along with embedded systems, small monitoring device or sensors can be implanted in the patient's body. These sensors can form a network, which is termed as the body sensor network (BSN).These networks are very helpful in continuous monitoring of patients. Cardiac disease can be easily diagnosed if the patient's ECG is monitored continuously. This is one of the common applications of BSN. For other critical diseases also, BSNs can be used.

1.1.2.7 Education

At present, internet access is easily available to a majority of the population around the world. Students and academicians around the world can now easily access each other's work. Internet has provided the world with a wide platform on which experts can deliver the sessions and brainstorming of ideas has become possible and fruitful. Beyond that, experts are asked to deliver lessons on the respective fields they are working on. With massive open online courses (MOOCs), majority of the population is able to receive online education. The idea of "flipped classrooms" is gaining popularity. In this, students will learn the concepts outside the classroom so that during the physical lecture innovative ideas can be discussed. The MOOCs are turning out to be a saviors for people in developing countries where many cannot afford the fees for higher education.

1.1.2.8 Smart Clothing

The wearable IoT is the next phase of IoT. The clothes will be designed for a special purpose with sensors embedded in them. For this, low-volume fabrics are preferred. However, in future conductive materials can be used in the production of cloth material. In order to make the clothes responsive, placement of sensors is very vital. The next important part is communication among the sensors. Lastly an interface needs to be designed in order to collect, store, and interpret the collected data. The tensile strength and the appearance of the cloth can be determined with the help of the collected information.

In recent years, various IoT projects have been developed for the benefit of society. A few prominent projects are discussed below:

1.1.2.9 Internet of Things-Architecture (IoT-A)

IoT-A is a European project aiming to develop a reference architecture model for the IoT with interconnected technologies, smart devices, resources, and services. IoT-A is a group of university and industry-funded institutions and experts in September 2010 by the European Union under the FP7 framework [5]. *"It can be seen as an umbrella term for interconnected technologies, devices, objects, and services."* The main challenge of IoT-A is to find a way to connect different devices, applications, and technologies in a single coherent network. To achieve this IoT-A provides a set of common protocols and interfaces and also the architecture to handle interoperability, heterogeneity, and scalability issues of IoT. IoT-A has three models: (i) domain model, (ii) information model, and (iii) communication model. Domain models define the IoT object attributes, that is, name and identity and define the relation between devices. Information model defines the methods to store and retrieve information and communication model defines the protocol stack for the IoT communication. Kramp and his team have developed IBM infrastructure platform for wireless sensor networks and published specifications for an embedded virtual-machine byte code, platform application programming interfaces (APIs), and protocols.

IoT-A follows the four-step process depicted in Figure 1.1. The process discusses the underlying assumptions, motivation, and how the existing IoT architecture can be used as a methodology. In the second step, IoT-A analyzes different business scenarios and stakeholders to understand the requirements of the IoT-A. A stakeholder analysis helps to understand the aspects of the architecture model needed for different stakeholders and their concerns. With the interoperability support from multiple organizations and IoT reference models, the new IoT-A reference model is proposed. IoT-A is the abstraction layer and understands communication between heterogeneous

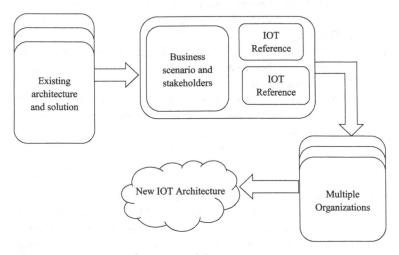

Figure 1.1   IoT: A reference model.

devices. IoT-A reference architecture focuses on abstraction sets of mechanisms other than application architecture.

1.1.2.10 Coordination and Support Action for Global RFID-Related Activities and Standardization (CASAGRAS)

Coordination and support action for global RFID-related activities and standardizations (CASAGRAS) was a project financed by the EU focused on foundational studies on international questions about RFID in support of IoT. It provided regulations and standardization of RFID [6]. CASAGRAS defines the components of IoT as a global network infrastructure, which is similar to the world-wide internet and allows messaging from one device to another. CASAGRAS has focused on the need for identification of each physical object in the IoT. It defines the virtual object, sensors, actuators, and connectivity. The virtual object is a representation of a proxy relationship with the physical objects. Sensors are devices that measure physical, chemical, and other forms of quantitative data. Actuators are used to control the system and are coupled with sensors. Connectivity is the interface between source data and a device. CASAGRAS architecture model depicted in Figure 1.2, consists of three layers as follows:

1. Physical layer
   In the physical layer, object and things are identified. The physical layer contributes functional components to the IoT, that is, technologies that connect objects and carry data like RFID, Zigbee, and Bluetooth.
2. Gateway layer
   In the gateway layer, there are layers that act as a mediator between connected devices and host management systems.
3. Application layer
   Application layer sends user information or functional data to the connected devices through gateway layer to support applications and services.

1.1.2.11 MAGNET and MAGNET Beyond

My personal Adaptive Global NET (MAGNET) was an EU-funded project in the area of personal networks. MAGNET has developed user-centric system and business model concepts for secure person network in the multi-network, multi-device and multi-user environment

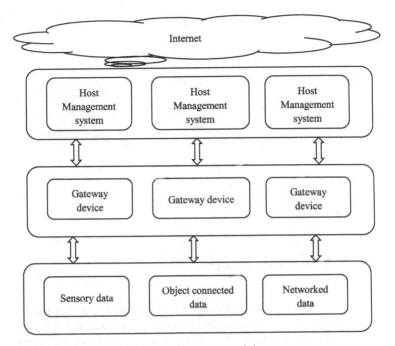

Figure 1.2 CASAGRAS architecture model.

[7]. MAGNET has 30 partners from 15 countries, which are highly acknowledged industrial partners, universities, and research centers. The objectives of MAGNET are to design and implement technologies and protocols that are needed to build a personal network according to the user's requirements in terms of quality and secure service. It implements a multilayer security concept to ensure the security and privacy of user's data. It designs efficient, flexible, and scalable air interface for the components of personal networks. It collaborates with other projects like MATRACE, 4MORE, PHONIX, WINNER, PULSERS, E2R, and AMBIENT NETWORK to implement user profiles, context and service discovery, PN to PN communication, trust and identity management, universal convergence layer, PN federation, and commercialization of PN architecture.

A personal network is a concept of supporting professional and private activities in an ad hoc network and infrastructure-based dynamic networks. Users can access personal devices at anytime from

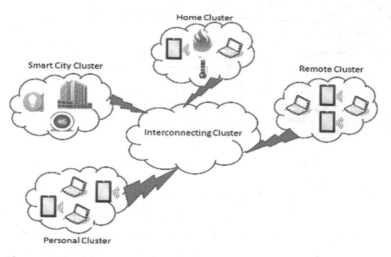

Figure 1.3 MAGNET architecture.

anywhere as depicted in the Figure 1.3. Configuration and connectivity of personal devices depends upon time and place and requires resources. MAGNET Beyond takes the PN concept further to form the PN federation (PN-F). Devices of two different PNs communicate with each other seamlessly and for a limited time to complete a specific task. The features of MAGNET features are listed below:

- MAGNET protocol is designed based on the Linux platform and IPV4 networking.
- Networking is built using universe convergence layer. The gateway node functionality was divided into two parts: the gateway and edge router.
- Security for MAGNET PN prototype is based on certified PN formation protocol and data is protected by encryption based on keys derived using the certificates.
- Service discovery is done by the service management platform, which provides context-aware service discovery for personal networks. Modified Universal Plug and Play (UPnP) device modules were responsible for discovery, registration, and control.
- MAGENT supports context management via secure context management framework. SCMF is responsible for storing, processing, and delivering context information.

Two applications developed to test MAGNET's real use are Icebreaker and Lifestyle Companion. The aim of the Icebreaker application was to demonstrate personal network capabilities at a big event like a conference. The user can utilize "Icebreaker service" offered by the conference organizers. Users can access user services to subscribe and check into the conference. The services also use the name, contact information, and business interest to access equipment in a showroom and to display presentations. The Lifestyle Companion application is an electronic personal trainer. When the user enters the gym, it asks the user to step onto a scale. The weight is taken and accordingly certain exercises are recommended.

1.1.2.12 BUTLER

BUTLER is the EU project with the aim of providing a set of guidelines and principles to build IoT systems [8]. BUTLER effectively performs integration of platforms and is extensible to add new functional components to the architecture. BUTLER facilitates to add value-added services such as localization, context capturing, and security management to build effective context-aware information systems, which seamlessly operate across various scenarios in the smart life environment. The major challenges are to provide privacy and security in heterogeneous systems. It provides a service-oriented architecture for dynamic, interoperability, and complex event processing.

BUTLER systems are deployed using three entities: smartObject, smartServer, and smartMobile. BUTLER architecture has identified four layers for all functionalities to fulfill requirements such as context management, communication, services and device management, depicted in Figure 1.4.

- The communication layer defines the end-to-end infrastructure that connects smartObject, smartMobile, and smartServers.
- Data/context management layer defines data models, interfaces, and procedure for data collection from data sources. It also helps to process the data in order to provide context information and transfer the raw data to high-level information.
- Device management layer defines the necessary components of management, maintenance of smartObject, services, configuration, software management, and diagnostics.
- Service layer defines interfaces for description, discovery, binding, deployment, and context-aware services.

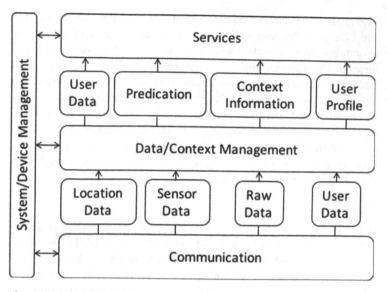

Figure 1.4   BUTLER context model.

BUTLER focuses on five scenarios as follows:

- Smart homes and offices
- Smart shopping
- Smart mobility and transport
- Smart healthcare and wellness; and
- Smart cities

1.1.2.13 Smarter Cities Data Management (SMARTIE)

SMARTIE is the EU project that aims to develop a framework for IoT to store, process, and distribute the big data for use in smart city applications [9]. SMARTIE framework provides security, privacy, and trust by embedding security mechanisms in the IoT framework instead of providing them as extra plug-ins. It is based on IoT-A architecture. SMARTIE components are organized into functional groups like IoT-A project. It has seven functional groups: application, management, service organization, virtual entity, IoT service, communication, and security functional groups. SMARTIE uses the Discovery framework for discovery of the IoT resources using Constrained Application Protocol (CoAP) and CoAP resource directory. Elliptic curve cryptography (ECC) based distributed capability access control (DcapBAC) mechanism is used for access control. DcapBAC uses XACML and JSON for lightweight approach, as IoT devices are resource constrained. SMARTIE project is demonstrated for few scenarios such as energy management, public transport, traffic management, and city information center.

1.2 IOT ARCHITECTURE: LAYERED PERSPECTIVE

IoT majorly consists of resource constraint devices. These devices have different communication, computational, and storage capabilities. The organization of these devices plays an important role in any IoT application. IoT architecture follows the layered approach [10] and 10C model, which is discussed in the next chapter. This layered approach has four layers, as depicted in Figure 1.5. In this approach, devices are at the bottom level, followed by the interconnection, service, and application layers in order.

- *Device Layer*: This layer consists of heterogeneous devices. These are heterogeneous in all sense, that is, size, communication

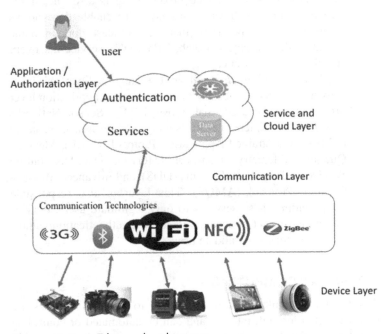

Figure 1.5 IoT layered architecture.

capability, energy source, computational capacity, identification, etc. In any IoT application, different devices may require to work together. Due to this heterogeneity, organizing and working of IoT application requires gateways that can communicate with all types of devices. Devices can communicate with the help of Wi-Fi, Zigbee, BLE, local area network (LAN), Long Range Wide Area Network (LoRaWAN), and many more.

- *Communication Layer*: This layer provides communication between users and IoT services provided by IoT devices. Users can access these services using the internet, hence addressing and identification of IoT devices is the basic challenge of this layer. Various protocols can be used in this layer, that is, IPv6 *Routing Protocol* (RPL), 6LowPAN, IPv4/IPv6, etc.
- *Service and Cloud Layer*: IoT services are developed based on information sensed by IoT devices. The information storage, information processing, and forming services are important challenges of this layer. In various IoT applications, Cloud computing is used for information storage and plays an important role in service utilization. Users should be made aware of these IoT services. Hence service broadcasting is also one of the important functionalities of this layer. To enable this various technologies can be used, that is, multicast domain name system (mDSN), physical web, UPnP, DNS service discovery (DNS-SD), HyperCat etc.
- *Application/Authorization Layer*: The functionality of this layer is to provide IoT services to the users. Various application layer protocols can be used, that is, MQTT For Sensor Networks (MQTT-SN), Extensible Messaging and Presence Protocol (XMPP), Constrained Application Protocol (CoAP), Message Queuing Telemetry Transport (MQTT), Data-Distribution Service for Real-Time Systems (DDS), and Advanced Message Queuing Protocol (AMQP). This layer is also responsible for providing these services to only legitimate users. Service owners have a set of access policies and authentication/authorization protocols to validate users.

1.3 SMART COMPUTING

Smart computing plays a crucial role in IoT. In IoT devices are interconnected through internet and can be automated or controlled remotely. Further, in smart computing, IoT devices can take action based on information gathered by the IoT network without human intervention as in many applications humans are unreachable. These actions can be taken based on learning techniques as traditional programming may be a very expensive solution. IoT with smart computing can enhance usefulness of IoT applications and can also increase revenue.

Functionalities of IoT with Smart Computing

- *Smart Home*: IoT devices in refrigerators can estimate quantities and expiry of food items on the basis of which notifications can be sent for the purchase of the items. It also recommends recipes based on the available food items. Sensors at the garden measures humidity and controls irrigation accordingly. Based on temperature inside the house the temperature of the air conditioner is managed.
- *Manufacturing Industries*: Classification of faults can be done based on sensor's diagnostic data. Further prediction of device failure can be done, which can be of help for device maintenance. Various mechanisms can be suggested for minimizing the energy consumption of battery-operated devices based on the energy consumption pattern.
- *Intelligent Transport*: The recommendation of the number of buses required for a particular route can be done based on the pattern of the crowd in public buses. Maintenance of vehicles can be scheduled based on sensor readings. Traffic signal timings can be changed dynamically based on the number of vehicles at a particular side.
- *Healthcare*: Alarms can be raised if there is any change in the regular pattern of a patient's body sensor data. Early detection of various diseases can be done by data analysis.

1.4 IOT DESIGN: ISSUES AND CHALLENGES

1.4.1 Design Issues

IoT applications are implemented based on the information provided by the IoT sensors. These data may be very large (Big Data) or small and may be kept on local servers or on the Cloud. These data are heterogeneous and can lead to various design issues for developing the IoT applications [11–16]. Few design issues of data are discussed below:

- *Volume*: Volume is defined as the amount of data that is being created and stored. It is increasing exponentially from terabytes to petabytes as days are passing. As the size is increasing, the storage capabilities are also making advancements. What cannot be captured in earlier days is possible now. This big data can be classified on the basis of the generated data (type of data) and time at which it is time generated. Type of data is often referred as the variety of data. Text data and audio data require different technologies for processing.
- *Velocity*: The rate at which data is being created is called as the velocity of the data. In legacy systems traditional data analytics was done on the updates received on a daily, weekly, or monthly basis. As real-time data is being generated in larger volumes, analyzing and processing this data is highly necessary in order to make informed decisions. Time is a very important factor. Some of the domains like retail, telecommunications, and finance are able to create high-frequency data. Data generated data from the mobile app can be useful in offering more personalized services to the user on the basis of user's preference.
- *Variety*: Variety refers to the heterogeneous nature of the data that is being created and stored. This data is categorized into three main sections:
  - Structured data
  - Semistructured data
  - Unstructured data

  If organization of the data can be done with the help of existing data models, then that data is called as structured data. The data that is represented with the help of rows and columns are excellent examples of structured data. Only 5 percent of the data is available in such a format. If data cannot be organized using any data model, it is known as unstructured data. Examples of such data include video, text, and audio data. Semistructured data lies in between structured and non-structured data. Extensible Markup Language (XML) is one of the best examples of semistructured data.
- *Veracity*: Unreliability associated with data sources is referred to as veracity. For example, sentiment analysis with the help of social media platforms is bound to create uncertainty in the results. Hence there is a need of a mechanism that will be used to manage the uncertain and unreliable nature of data.

- *Variability*: Inconsistency in velocity of the big data causes flux in the flow rate of data, which is called as the variability of the data. Data is being created from a number of sources and management of such diversified data is really a big task. As data is being generated from diversified resources, it has different underlying semantics.
- *Low-value density*: Raw data is not usable. Data needs to be processed and analyzed in order to make sense out of it. Logs from general stores can be analyzed to make new formation of products. It will in turn help in fetching more business and revenue. It can also be used to analyze a customer's behavior.

1.4.2 Challenges

IoT has numerous challenges such as standardization, communication and connectivity interface, and heterogeneity [17–25]. Some of the challenges are addressed below.

- *Mobility*: An IoT system is a collection of different devices working in collaboration with each other. Every device has a goal associated with it. For example, sensing leak in irrigation system or a failure in AC unit. Collaboration requires more complex functionalities. The devices are categorized into two parts. The first is static and second is mobile. Organization of IoT devices will enable the system to be designed with complex abilities. The static devices are more manageable to work with as compared to mobile devices.
- *Heterogeneity*: Heterogeneity is the base of IoT devices. These devices vary in size, computational power, their capabilities, interface for communication, power source, and a whole bunch of parameters. To design an IoT service while working with such heterogeneity is a real challenge.
- *Scalability*: The use of IoT devices in our day-to-day life is increasing with time. It will continue to grow exponentially in the future. IoT applications' basic requirement is to have a number of heterogeneous devices (those can static or mobile). According to the application requirement, the number of devices involved can change. To support such a type of heterogeneity, IoT needs to provide scalability and support.
- *Energy efficiency*: IoT systems are comprised of a number of devices. All the devices operate on some kind of a battery

source. These devices require power in order to sense data, transmit data, process of data, and to take action if necessary. IoT systems should be designed to be more energy efficient so as to sustain the life of devices.

- *Security*: The IoT systems are extremely vulnerable to attacks because of the following reasons:
  - Most of the devices are in the open and unguarded; hence it is very easy to manipulate them.
  - Wireless communication happens between these devices. It is more vulnerable to attacks and very simple to compromise.
  - As most of the devices are operated on a battery source, complex techniques cannot be implemented even for security purposes.

- *Privacy and context management*: The service providers at present operate on "take it or leave it" strategy. It means if you want a particular service, there are policies and conditions that are designed by the service providers for sharing personal data. If you want the service, you have to accept these terms. These terms lead to the invasion of privacy for a typical user. Hence sharing of the personal data with a number of service providers will surely result in the violation of privacy for the user. Hence there is a need of IoT applications to figure out what the user wants with minimal amount of data shared by the user.

- *Context management and knowledge retrieval*: The basis of any IoT application is contextual information. After acquiring the information, knowledge retrieval is the next main task. But both these tasks face challenges because of the heterogeneity of contextual information.

REFERENCES

1. Gubbi, J., Buyya, R., Marusic, S., and Palaniswami, M. (2013). Internet of Things (IoT): A vision, architectural elements, and future directions. *Future Generation Computer Systems*, 29(7), 1645–1660.

2. Lee, I., and Lee, K. (2015). The Internet of Things (IoT): Applications, investments, and challenges for enterprises. *Business Horizons*, 58(4), 431–440.

3. Serbanati, A., Medaglia, C. M., Ceipidor, U. B., and Turcu, C. (2011). Building blocks of the Internet of Things: State of the art

and beyond. *Deploying RFID-Challenges, Solutions, and Open Issues*, 251–366.

4. Shinde, G.R. (2017). Cluster Framework for Internet of People, Things and Services.

5. S. Meissner, S. SAP, and T. (2012) NEC, "Internet-of-Things Architecture IoT-A Project Deliverable D2. 5-Adaptive, fault-tolerant orchestration of distributed IoT service interactions".

6. I. G. Smith, Anthony Furness, Ken Sakamura, Ricky Ma, Eldor Walk, Craig Harmon, Patrick Guillemin, Yong-Woon Kim, Paul Chartier, David Armstrong, Y.-W. Kim, P. Consultants – Paul Chartier, and R. Ltd -David Armstrong, "2 CASAGRAS – Final Report The CASAGRAS partnership CASAGRAS (Grant Agreement 216803) is a Coordination and Support Action for Global RFID-related Standardisation Activities involving, in particular, organisations from." Eu Fp7 Project Casagras, "Casagras Final Report: RFID and the Inclusive Model for the Internet of Things," 2009.

7. R. Prasad, Ed., *My Personal Adaptive Global NET (MAGNET)*. Dordrecht: Springer Netherlands, 2010.

8. "Project Deliverable Project Number: Project Acronym: Project Title: Integrated Project Internet of Things Title D3.2 Integrated System Architecture and Initial Pervasive BUTLER proof of concept," 2013.

9. P. Langendörfer and J.-M. Bohli, "At A Glance: SMARTIE Secure and sMArter ciTIes data management Project Coordinator SMARTIE: Secure and smarter cities data management," 2013.

10. M. Westerlund, S. Leminen, and M. Rajahonka (2014). Designing business models for the Internet of Things, *Technology Innovation and Management Review*, 4(7), 5–14.

11. Labrinidis, A., and Jagadish, H. V. (2012). Challenges and opportunities with big data. *Proceedings of the VLDB Endowment*, 5(12), 2032–2033.

12. Sivarajah, U., Kamal, M. M., Irani, Z., and Weerakkody, V. (2017). Critical analysis of Big Data challenges and analytical methods. *Journal of Business Research*, 70, 263–286.

13. Bhadani, A., Jothimani, D. (2016), Big data: Challenges, opportunities and realities, In Singh, M.K., and Kumar, D.G. (Eds.), *Effective Big Data Management and Opportunities for Implementation* (pp. 1–24), Pennsylvania, PA: IGI Global.

14. Nasser, T., and Tariq, R. S. (2015). Big data challenges. *Journal of Computer Engineering & Information Technology*, 4, 3. doi: http://dx. doi. org/10.4172/2324, 9307(2).

15. Jagadish, H. V., Gehrke, J., Labrinidis, A., Papakonstantinou, Y., Patel, J. M., Ramakrishnan, R., and Shahabi, C. (2014). Big data and its technical challenges. *Communications of the ACM*, *57*(7), 86–94.

16. Jaseena, K. U., and David, J. M. (2014). Issues, challenges, and solutions: big data mining. *CS & IT-CSCP*, *4*(13), 131–140.

17. Rose, K., Eldridge, S., and Chapin, L. (2015). The internet of things: An overview understanding the issues and challenges of a more connected world. The Internet Society (ISOC), 22.

18. Tamane, S., Solanki, V. K., and Dey, N. (Eds.). (2017). *Privacy and Security Policies in Big Data*. Pennsylvania, PA: IGI Global.

19. Vimal, S., Khari, M., Crespo, R. G., Kalaivani, L., Dey, N., and Kaliappan, M. (2020). Energy enhancement using Multiobjective Ant colony optimization with Double Q learning algorithm for IoT based cognitive radio networks. *Computer Communications*, *154*, 481–490.

20. Dey, N., Fong, S., Song, W., and Cho, K. (2017, August). Forecasting energy consumption from smart home sensor network by deep learning. In International Conference on Smart Trends for Information Technology and Computer Communications (pp. 255–265). Springer: Singapore.

21. Mukherjee, A., and Dey, N. (2019). *Smart Computing with Open Source Platforms*. Boca Raton, FL: CRC Press.

22. Mukherjee, A., Panja, A. K., and Dey, N. (2020). *A Beginner's Guide to Data Agglomeration and Intelligent Sensing*. London: Academic Press.

23. Mhetre, N. A., Deshpande, A. V., and Mahalle, P. N. (2016). Trust management model based on fuzzy approach for ubiquitous computing. *International Journal of Ambient Computing and Intelligence* (IJACI), 7(2): 33–46.

24. Babar, S., Mahalle, P., Stango, A., Prasad, N., and Prasad, R. (2010, July). Proposed security model and threat taxonomy for the Internet of Things (IoT). In International Conference on Network Security and Applications (pp. 420–429). Springer: Berlin.

25. Mahalle, P. N., Anggorojati, B., Prasad, N. R., and Prasad, R. (2013). Identity authentication and capability based access control (iacac) for the internet of things. *Journal of Cyber Security and Mobility*, 1(4): 309–348.

2

INTERNET OF THINGS
APPLICATION DEVELOPMENT

2.1 APPLICATION DEVELOPMENT PHASES

Due to rapid advancements in e-commerce and m-commerce as well availability of internet at a faster and cheaper rate, the use of mobile applications and IoT-enabled applications is increasing at a quicker rate. IoT solutions have become an integral part of everyday life and IoT application development has vast becoming potential in business. Smartphones, sensors, cameras, and RFID are the major components in all IoT use cases. Minimum human intervention for exchange of information and interconnecting devices using wired and wireless communication medium are the key steps in building IoT applications.

2.1.1 Principles

In this section, a brief description of the main principles to be taken into consideration before developing IoT applications is given. These principles cover issues like secure and real-time data collection, data management, IoT platforms, and IoT–Cloud convergence and they are discussed below:

1. *Secure data collection*: In many IoT applications, data is collected from a very large number of sensors that are deployed in a closed environment or also distributed over several locations. In such cases, data collection in a secure way becomes a major issue. Necessary mechanisms for ensuring data integrity and lightweight encryption for sensitive data must be adapted.

2. *Real-time data*: In all scalable IoT applications, data is collected in real time for processing and therefore traditional packet transfer mechanism for data transfer is inadequate. For efficient streaming, efficient techniques must be adapted for collecting real-time data.

3. *IoT platforms*: Data received from IoT devices is quite huge and it is collected from heterogeneous devices. To store and process this data, an appropriate IoT platform must be selected in advance.

4. *Cloud-centric development*: Centralized storage and fast delivery of responses to the user are main characteristics of IoT applications. Cloud-centric application development is more appropriate in such cases, which also optimizes the cost and is also suited for carrier networks.

5. *Data management*: Due to rapid advancement in semiconductor technology, space is getting cheaper with time. The main challenge is how to make sense out of this big data. Data science and machine learning techniques must be adapted for effective data management.

2.1.2 C Model

This section presents IoT application development phases using a novel C model. With reference to the layered perspective of IoT architecture presented in Chapter 1, there are mainly three layers in IoT: device layer, access layer, and application layer. The 10 C model proposed in this book will help the reader understand various phases and operational steps to be taken care during IoT application development.

(a) *Category*

Before we start developing any IoT application, it is important to decide whether the underlying IoT application is representing an indoor or outdoor use case, as there are different design issues for these use cases and they also have different sets of requirement and challenges.

(b) *Component*

The finalization of category and requirements of the use case helps to decide the components that are required for development. The major IoT components include smartphones, RFID, sensors, cameras, and various platforms that are required to process the captured data.

(c) *Cost*

Cost is an important factor from the business perspective and therefore it is better to estimate the cost in the early phases of development. Types of components, their configurations, and number of units of each component help to estimate the cost.

(d) *Connectivity*

Establishing connectivity between different devices is the first and important step in IoT application development. Devices could be connected to the user using different communication technologies such as Ethernet, WiFi, Bluetooth, WLAN, or Zigbee. These technologies differ in communication range, energy requirement, data rates, and interface.

(e) *Communication*

Once the devices are connected, communication is initiated between the devices. Communication can take place in multiple forms like between two devices, between the device and user, between two users, and between the user and device.

(f) *Cooperation*

For enabling the operation of various IoT services, multiple sensors are used for communication. This multi-device communication helps to initiate cooperation and is required for service discovery, device discovery, and availing services from them.

(g) *Convergence*

Devices can use multiple communication technologies, resulting in convergence. For example, when a device is in an indoor scenario, then it is connected to the internet using WiFi and in outdoor use cases, 4G or 5G cellular network is used.

(h) *Content*

Every device connected to the internet generates some data and posts it on the Cloud. There are many plug-and-play devices that can download data from the Cloud. These data-centric operations are periodical or real time and are based on the type of application.

(i) *Context*

Context plays an important role in ubiquitous applications. The context of a use case can be decided based on the content generated from the application. For example, data from applications like health care, agriculture, and smart home can be used to easily decide the context and related information.

(j) *Complexity*

There is complexity associated with IoT applications and it is applicable to all aforementioned phases in application development. Time, space, and skill requirement also decide the complexity of the application.

2.2 WIRELESS TECHNOLOGIES FOR IOT

Wireless technologies play a key role in all IoT use cases. In indoor use cases, all devices are connected to the gateway device using a variety of options, which include Bluetooth, Zigbee, or WiFi. In outdoor use cases, options are WiFi or 2.5G/3G/4G.

2.2.1 Bluetooth, Bluetooth Low Energy, and Bluetooth 5

Bluetooth uses 2.4 GHz of operating frequency available in Industrial, Scientific and Medical (ISM) band and is mainly used to initiate communication between two devices. The operational range of Bluetooth is 2,400 to 2,483.5 MHz globally working with 79 channels where each channel has a range of 1 MHz. The maximum range offered by Bluetooth is 10 meters and it uses a client–server model to enable data transfer in full duplex mode at the data rate of 64 kbps [1]. A Piconet network of devices is used for Bluetooth communication supporting seven devices where one is a master and others are slaves.

However, current consumption in Bluetooth is 30 mA and to overcome this limitation, all latest devices are using Bluetooth Low Energy (BLE) where the current consumption is 15 mA. The manufacturer interoperability, more range for communication, low-cost implementation, and lightweight suite are main advantages of BLE. BLE is more suitable for industrial applications where the communication range is 250 meters. BLE is not backward compatible with the devices supporting Bluetooth.

The emerging prominent standard of Bluetooth more suited for IoT application development is Bluetooth 5. It is revised and customized based on the design issues and requirements for IoT use cases offering a data rate of 2 Mbps, which is twice that of BLE. The transmission power level in Bluetooth 5 is enhanced to +20 dBm with increased advertised capacity, which is eight times more than BLE, making it more suitable to create interactive IoT applications.

2.2.2 Zigbee

Zigbee is built on the top of IEEE 802.15.4 and is mainly used in the wireless personal area network (WPAN) [2]. It consists of two key layers: (i) physical layer and (ii) medium access control sublayer. Router, end device, and coordinator are three main actors in Zigbee. The coordinator acts as a responsible device for data transmission activities and storage. End devices are generally tiny devices with resource constraints like limited power, memory, and computational power and have the ability to communicate with the parent device. Zigbee is suitable for indoor use cases like smart home or smart shop for monitoring and controlling purposes. Zigbee operates at frequencies of 868 MHz, 902–928 MHz, and 2.4 GHz and the data rate for communication is 250 kbps. The typical communication range is 30 meters for indoor and 100 meters line of site for outdoor applications. Each Zigbee network can have 64Knodes.

2.2.3 WiFi

WiFi is a widely used wireless communication technology used in both indoor and outdoor use cases. It is based on 802.11, which represents the widely known standards for wireless local area networks. Access points or hotspots are used to connect devices to the internet providing a data rate of 54 Mbps. The operating frequencies used by WiFi is 2.4 and 5 GHz in ISM band and provides communication range of 20 meters and 100 meters for indoor and outdoor use cases, respectively. It provides interoperability for various WiFi devices, is cost effective, and integrated with all widely used devices ranging from laptops to TV. WiFi functions in star topology where internet gateway is the center node and consists of a physical layer and medium access control layer [3]. The main specifications of 802.11 are 802.11a (operating frequency: 5 GHz, speed: 54 Mbps), 802.11b (operating frequency: 2.4 GHz, speed: 11 Mbps), and 802.11g (operating frequency: 2.4 GHz, speed: 20 Mbps).

2.2.4 6LowPAN

Due to the emergence of IoT-enabled applications where resource constraints are the main challenge, low-power and lossy networks become a good candidate. The objective of these networks is to save energy by managing other constraints like memory and duty cycle. IPv6 over low-power wireless personal area network (6LoWPAN) [4] is a key option to meet the requirements of resource constraints. Energy saving can be carried out by optimizing the amount of data transmission. In resource-constrained networks like IoT, requirement of data transmission is less than 1280 bytes, which is the transmission rate of IPV6.These networks use IEEE 802.15.4 physical layer where the rate is 178 bytes per frame. In view of this, 6LoWPAN header compression is required using IPV6 packets over 802.15.4.

2.2.5 3G/4G/5G

The internet is getting faster and cost is becoming cheaper gradually, the credit for which goes to advancements in broadband technology. 3G/4G/5G represent the generations of wireless technology that are related to the speed of internet and are dependent on the incoming signal strength. This wireless communication technology is mainly used in outdoor IoT use cases and 2G is the first generation in this family, which is almost obsolete now. The detailed specifications of these generations are as follows [5]:

    3G: Introduced in 2001, this generation is built on universal
        mobile telecommunication system as the core network back-
        bone. It provides speed up to 14 Mbps and operating range is
        2,100 MHz. It has broadband capabilities with a bandwidth
        capacity of 15–20 MHz.
    4G: Introduced in 2009, it has an enhanced data rate that
        differentiates 4G from 3G and is mainly based on technologies
        like multiple input multiple output and orthogonal frequency
        division multiplexing. The main 4G standard is WiMAX. 4G
        supports a top speed up to 1 Gbps and an average speed up to 50
        Mbps. Due to these enhancements, 4G can support multimedia,
        voice, and interactive communication. 4G is mainly used in ad
        hoc, multi-hop, and scalable mobile networks.
    5G: Introduced in 2018, 5G operates in the range of 30–300
        GHz with low latency and broader coverage due to small cells

operated in radio millimeter band. The internet service provided by 5G is World Wide Web and supports high-end applications like real-time streaming of videos.

2.2.6 LoRAWAN

Low-power wide area network (LPWAN) [6] uses LoRaWAN protocol, which support requirements for IoT applications with mobility, less power, and two-way communication. An improved network capacity and optimized energy consumption are key features of LoRaWAN supporting a data rate of 0.3–50 kbps. LoRaWAN operates in the band of 2.4–5 GHz and uses star topology. For IoT applications, this protocol is also supported by three types of unique keys: network, application, and device keys for security purposes.

2.2.7 Sigfox

Sigfox is a unique service provider solution for IoT applications offering software-based Cloud-centric communication solutions with optimized energy consumption and less cost. Lightweight protocol, small payload, star topology, and ultra-narrow band radio modulations make this solution a better choice for IoT applications. It is also reported that Sigfox is more scalable as compared to other solutions [7].

2.3 PROTOCOL STACK

Different working groups such as the Internet Engineering Task Force (IETF), World Wide Web Consortium (W3C), Institute of Electrical and Electronics Engineers (IEEE), EPCglobal, and the European Telecommunications Standards Institute (ETSI) provide diversified protocols for each layer of the IoT as depicted in Figure 2.1 such as the device/physical layer, communication layer, service layer, and application layer. In this section, Application Layer protocols for IoT are discussed as highlighted in Figure 2.1.

2.3.1 CoAP

Constrained application protocol (CoAP) is an application layer protocol developed by IETF intended to be used in resource-constrained IoT devices [8]. It is designed for "machine to machine communication" type of applications such as automated smart

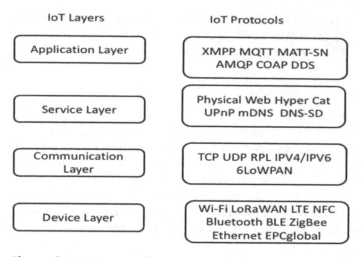

IoT Layers	IoT Protocols
Application Layer	XMPP MQTT MATT-SN AMQP COAP DDS
Service Layer	Physical Web Hyper Cat UPnP mDNS DNS-SD
Communication Layer	TCP UDP RPL IPV4/IPV6 6LoWPAN
Device Layer	Wi-Fi LoRaWAN LTE NFC Bluetooth BLE ZigBee Ethernet EPCglobal

Figure 2.1 IoT protocols.

building and automated service center. CoAP runs on devices that support User Datagram Protocol (UDP) communication. It is designed for simplicity, low overhead, and multicast support in resource-constrained environments. CoAP is a lightweight protocol, which is intended to be used over HTTP similar to the representational state transfer (REST). It provides simple request/response interaction model between application devices. It uses the Uniform Resource Identifier (URI) to identify a resource. CoAP has a two-layered architecture – a bottom layer designed to deal with UDP data communication and asynchronous message communication. This layer uses back off exponential for reliable data communication and also detects message duplication. The upper layer deals with request/response communication over HTTP. CoAP consists of four types of messages – non-confirmable, confirmable, reset (RST), and acknowledgment (ACK) – and four types of working modes: confirmable, non-confirmable, piggyback, and separate mode as represented in Figure 2.2. In the confirmable mode, the server sends ACK to the client after receiving the message. In the non-confirmable mode, ACK is not sent from server side, that is, unreliable communication. In the piggyback mode, ACK is sent with next message and in a separate mode ACK is sent by the server as a separate packet and the server needs to take fix time to send an ACK. Hence the client waits for ACK before sending the next message. Security is provided using Datagram Transport Layer Security (DTLS). There are three main components

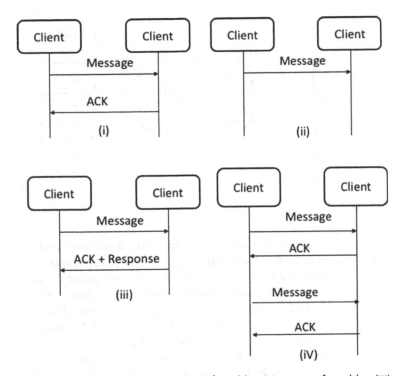

Figure 2.2 CoAP modes: (i) confirmable, (ii) non-confirmable, (iii) piggyback, and (iv) separate.

while considering security for CoAP: integration, authentication, and confidentiality. DTLS also avoids cryptographic overhead problems that occur in low-layer protocols. In the literature, CoAP is used in many IoT applications [9][10].

2.3.2 MQTT

Message queuing telemetry transport (MQTT) is a publish/subscribe lightweight messaging protocol [11]. It is used in applications where devices are power constrained and low network bandwidth is available. It is designed for unreliable networks and high-latency devices. It was designed to handle a large amount of remote measurement data in low-bandwidth networks. MQTT has three components: publisher, subscriber, and a broker, as depicted in Figure 2.3. MQTT is a publish/subscribe model type of protocol, where every device, for example, temperature sensor, humidity sensor, or the light sensor is a publisher.

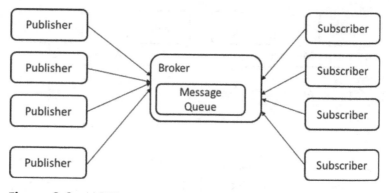

Figure 2.3   MQTT.

A publisher always establishes a network connection with a broker. The publisher and subscriber communicate to each other over TCP/IP network. Publishers are connected to the subscriber through a broker. Every publisher either publishes the message to the broker or subscribes to the messages from the broker. The data carried over the MQTT protocol is called Application message. The label attached to the application message is called Topic. The broker receives all messages from the publisher and then it filters out the messages and sends them to the interested subscriber based on "topic" name. MQTT can add additional security by adding encryption to the application data. We can pass a user name and password to the MQTT protocol for security. MQTT is used in many real-time IoT applications such as intelligent parking [12] and fire alarm [13].

2.3.3 AMQP

Advanced message queuing protocol (AMQP) is an open standard message-oriented, publish/subscribe, reliable, and secure protocol for "machine to machine" (M2M) communication. AMQP runs over the transport layer protocol. AMQP works similar to the email server and follows the store and forward architecture [14].

In AMQP, the publisher sends a message to the subscriber through a broker. The broker is divided into two parts: exchange and message queue, as depicted in Figure 2.4. The exchange has the same function as a transfer agent in the email server and message queue as a mailbox. The exchange is responsible for receiving published messages by different publishers and to distribute the received published messages

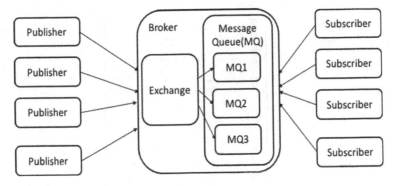

Figure 2.4   AMQP.

to the respective queues based on roles and conditions of each subscriber. The message queue stores the messages on a disk. The subscriber accesses messages from the message queue.

2.3.4 XMPP

Extensible messaging and presence protocol (XMPP) is an open and XML-based protocol [15]. It was designed for chatting and sending small instant messages that are delivered immediately to the user. It uses both client/server and publish/subscribe architecture, wherein it wholly depends upon the application to choose suitable architecture for its application. It is designed for real-time applications and it supports small memory footprint as well as low-latency message exchange. The XMPP architecture is similar to an email server, where anyone can run their own XMPP server enabling to take individual control. XMPP is secure using simple authentication and security layer (SASL) for authentication and transport layer security (TLS) for a secure channel. It adds end-to-end encryption to raise the security bar even further. XMPP runs over TCP/IP network. An XML message can create a lot of overhead to the network due to the headers, which increases power consumption, making it unsuitable for IoT applications.

2.3.5 Comparison of Application Layer Protocols

In the literature, AMQP is compared with MQTT by Luzuriaga et al. [16]. Use of AMQP protocol is recommended for reliable and scalable

TABLE 2.1
A Comparison of Application Layer Protocols

Protocol	Restful	Transport	Security	Request/Response	Architecture	Communication
CoAP	Yes	UDP	DTLS	Yes	Tree	D- to- D
	Yes	TCP	SSL	No	Tree	D to D
						D- to- C
AMQP	Yes	TCP	SSL	No	Tree	D- to- D
						D- to- C
						C- to- C
XMPP	Yes	TCP	SSL	Yes	Client–server	D- to- D
						D- to- C

D- to-D: Device to Device, D-to-C: Device to Cloud, C-to-C: Cloud to Cloud.

messaging over WLAN. In Table 2.1, the application layer protocols are compared and evaluated based on the parameters like transport layer protocol used, architecture followed, type of communication supported, security protocol used, that is, Datagram Transport Layer Security (DTLS) and Secure Sockets Layer (SSL). Different application protocols may be suitable for different IoT scenarios. Hence, it is not feasible to recommend a single protocol for all IoT applications. In the research work, Representational State Transfer (RESTful) approach is applied for usage of IoT services.

2.4 ELECTRONIC PLATFORMS

As mentioned earlier, every IoT use case collects data from multiple sensors and there is a need for electronic platforms to process this data before it is forwarded to the gateway device for further processing. This section provides a brief description of different electronic platforms and also a comparison of these platforms [17].

2.4.1 Arduino

Arduino is an 8-bit microcontroller board based on the ATmega3280P chip from the Microchip family. Arduino is an open source device platform used for building small electronics boards. Arduino supports software programming languages like C, C++, Python, and JavaScript. Arduino hardware board is available for a price of $20. The typical Flash code size is 32 kb and the RAM size is 2 kb. The power consumption of an Arduino is typically 20 mA. Arduino operates on 16 MHz clock frequency. Arduino does not have enough memory to build large embedded systems. Arduino boards are suitable for the projects where cost and power consumption are the primary factors. Different Arduino hardware boards are available with varying numbers of General-Purpose Input/Output (GPIO) pins and communication media such as Bluetooth, WiFi, and Controller Area Network (CAN).

Arduino is used in small projects like automatic room temperature control using temperature and humidity sensors, detecting motion, and controlling an automated soldering robot. Arduino boards are used to read various sensor data and controlling relay-based action or sending sensor data to the Cloud using the WiFi or Bluetooth module. There are various applications that can be implemented on Arduino boards. Typical IoT Applications on Arduino boards are automatic watering to home plants, door open and closing from internet, home automation, and motor controlling applications. Arduino is a microcontroller-based platform that does not have an RTOS system installed. Arduino is not suitable for large projects, where the real-time operating system is required, or large data processing and huge data storage are required.

2.4.2 Raspberry Pi

Raspberry Pi is a 32-bit microcontroller board based on the ARM1176JZF-S architecture. It consists of BCM2835 system on chip (SOC) from the Broadcom family and it is a multimedia chip. A low-cost single-board computer (SBC) board is used in Raspberry Pi and it is mounted on a Linux operating system. It uses an open source operating system called Raspbian, which runs a pool of open source software and is available for a price of $35.

The typical onboard storage for Raspberry Pi board is SD, MMC, and SDIO card slot and 256 Mb SDRAM memory. The power consumption of the Raspberry Pi board is typically 700 mA. Raspberry Pi operates on 700 MHz clock frequency. Raspberry Pi board does have Ethernet, USB 2.0, High Definition Multimedia Interface (HDMI) video output, audio output, and dual-core multimedia coprocessor. Raspberry Pi boards are tiny and portable in size, making them suitable for projects based on embedded architecture. They are also a good option for projects where high interactivity is a priority. In the academic field, Raspberry Pi is more popular for designing innovative projects in learning, bringing novelty in content delivery, and schooling projects. While Gbit Ethernet and WiFi connectivity make the Pi an ideal solution in an industrial IoT environment, the Raspberry Pi board is used in IoT-based intensive care unit (ICU) patient monitoring systems, automatic door control based on lighting, industrial automation, motion video, and a Web server.

One of the limitations of Raspberry Pi models is that they don't have a built-in real-time clock. Raspberry Pi board lacks interfaces including RS-485/232 serial ports, CAN, and economical 1-wire buses, and digital and analog inputs and outputs. The Raspberry Pi board being used in harsh industrial environments with high temperatures and strong vibrations leads to performance losses and damage to the board. The microSD cards can only retain a limited amount of information, making them unsuitable for industrial applications involving the storage of large amounts of data.

2.4.3 BeagleBone

BeagleBone Black is a 32-bit microcontroller board and uses ARM Cortes A8 as its key component in the architecture. It consists of Sitara AM335x SOC processor from the Texas family. The key characteristics of this board are less cost and they support development platform, which is community-supported. These boards are useful for low-power applications and supports open source software development. BeagleBone Black board is available for a price of $50.

The typical on board storage for the BeagleBone Black board is 4-Gb E-MMCS and 512 Mb DDR3 RAM memory. The power consumption of the BeagleBone board is typically 500 mA. BeagleBone Black operates on a 1 GHz clock frequency. BeagleBone black board does have a 3D graphics accelerator, neon floating point accelerator, fast Ethernet, USB host, micro-HDMI video output, and audio output and PowerVR SGX530 coprocessor. BeagleBone is preferred by beginners due to its simplicity, size, and is more suited for designing and implementing IoT applications. Other key characteristics of BeagleBone compared to Raspberry Pi are simpler networking options, compatible with Web browsers and USB cables and are more suited to projects that require taking inputs from external sources like sensors and actuators. Use cases for this board includes light system and smart irrigation system.

The limitations of Beagle Board includes not suitable for complex multimedia and Linux-based projects due to unavailability of the graphical and audio capabilities that the Raspberry Pi offers. BeagleBone Black does not have a larger community support. BeagleBone Black is a much costlier selection for add-ons and peripherals. Table 2.2 presents the comparison of various development boards.

TABLE 2.2
Comparison of Development Boards

S. No	Parameter	Arduino	Raspberry Pi	BeagleBone
1	Example controller	8 bit	32 bit	32 bit
2	Flash code	32 kb	SD, MMC card	4GB e-MMC
3	RAM size	2 kb	256 Mb	512 Mb
4	Microcontroller chip	Atmega3280P	BCM2835	AM335x
5	Clock	16 MHz	700 MHz	1 GHz
6	Operating system	None	Linux	Linux, Android
7	Price	20$	$35	$50

REFERENCES

1. Snellma Henrik, Savolainen Mikko, Knaappila Jere, and Rahikkala Pasi, Bluetooth ® 5,Refined for the IoT. Available online: www.silabs.com/documents/public/white-papers/bluetooth-5-refined-for-the-IoT.pdf, 2019.
2. Zigbee Alliance, Zigbee Specification. Available online: https://people.ece.cornell.edu/land/courses/ece4760/FinalProjects/s2011/kjb79_ajm232/pmeter/ZigBee Specification.pdf, 2008.
3. EnGenius Technologies, Wi-Fi Beacon Frames Simplified. Available Online: www.engeniustech.com/wi-fi-beacon-frames-simplified/, 2017.
4. RFC 8025: IPv6 over Low-Power Wireless Personal Area Network (6LoWPAN). Available online: https://tools.ietf.org/html/rfc8025.
5. R. Zeqiri, F. Idrizi, and H. Halimi, 2019. "Comparison of Algorithms and Technologies 2G, 3G, 4G and 5G," 3rd International Symposium on Multidisciplinary Studies and Innovative Technologies (ISMSIT), Ankara, Turkey, pp. 1–4.
6. LoRaWAN Alliance. Available online: https://lora-alliance.org/about-lorawan.
7. A. Lavric, A. I. Petrariu, and V. Popa. "Long Range SigFox Communication Protocol Scalability Analysis Under Large-Scale, High-Density Conditions," IEEE Access, 7, 35816–35825, 2019.
8. R. Jain, "Constrained Application Protocol for Internet of Things," *International Journal of Engineering and Technology*, 857, 1–12, 2014.
9. C. Cho, J. Kim, Y. Joo, and J. Shin, "An Approach for CoAP-based Notification Service in IoT Environment." International Conference on Information and Communication Technology Convergence (ICTC), Jeju Island, Korea, 2016, pp. 440–445.
10. D. Ugrenovic and G. Gardasevic, "CoAP protocol for Web-based monitoring in IoT healthcare applications," 23rd Telecommunications Forum Telfor (TELFOR), Belgrade, Serbia, 2015, pp. 79–82.
11. "MQTT." Available online: http://mqtt.org/.
12. P. Dhar and P. Gupta, "Intelligent Parking Cloud Services Based on IoT Using MQTT Protocol," in International Conference on Automatic Control and Dynamic Optimization Techniques (ICACDOT), Pune, India, 2016, pp. 30–34.
13. D.-H. Kang et al., "Room Temperature Control and Fire Alarm/Suppression IoT Service Using MQTT on AWS," in International Conference on Platform Technology and Service (PlatCon), 2017, pp. 1–5.

14. C. Trieloff et al., "AMQP Advanced Message Queuing Protocol Specification License."
15. "XMPP | An Overview of XMPP." Available online: https://xmpp. org/about/technology-overview.
16. J. E. Luzuriaga, M. Perez, P. Boronat, J. C. Cano, C. Calafate, and P. Manzoni, "A Comparative Evaluation of AMQP and MQTT Protocols Over Unstable and Mobile Networks," in 12th Annual IEEE Consumer Communications and Networking Conference (CCNC), Las Vegas, NV, 2015, pp. 931–936.
17. D.G. Costa and C. Duran-Faundez, Open-Source Electronics Platforms as Enabling Technologies for Smart Cities: Recent Developments and Perspectives. *Electronics* 7, 404, 2018.

3

IOT–CLOUD CONVERGENCE

3.1 INTRODUCTION

Chapters 1 and 2 have given us a vivid picture about IoT and the associated design issues and challenges. The chapters have also given us an insight into the IoT application development process, key technologies used, and the platforms available that support in the actual development of the applications. They have built a strong foundation for the rest of chapters in this book in which we will discuss about the importance of Cloud in IoT and the various approaches and techniques that can be applied in real-time scenarios encountered when developing an IoT application.

In this chapter, we primarily discuss how Cloud plays a vital role in realizing IoT to its full potential. The chapter is organized as follows. Section 3.1 discusses the opportunities and challenges that are encountered when the IoT–Cloud convergence comes into picture. Section 3.2 provides a bird's eye view on the architecture for convergence. Section 3.3 elaborates on data offloading and computing from the IoT perspective. Section 3.4 explains the concept of dynamic resource provisioning for IoT. Section 3.5 introduces the various test beds and technologies with respect to IoT–Cloud Convergence.

The topic under discussion being IoT–Cloud Convergence cannot be well explained without providing an insight into the basics of the Cloud technology in brief. The National Institute of Standards and Technology (NIST) defines Cloud computing as "a model for enabling ubiquitous, convenient, on-demand network access to a shared pool of configurable computing resources (e.g., networks, servers, storage, applications, and services) that can be rapidly provisioned and released with minimal management effort or service provider interaction" [1]. In simple terms, Cloud computing is on-demand delivery of scalable computing resources (storage, servers, etc.) provided as a service to the users over the internet, thereby allowing flexibility and efficient scaling according to one's business needs. As the Cloud computing model works on the 'pay-as-you-go' model (pay for only as much as you use), it helps greatly in reducing operational and infrastructure maintenance costs. Also, as the infrastructure setup and maintenance are taken care of by the Cloud service providers (Amazon, Google, Azure etc.) businesses can now concentrate on developing innovative products and applications pertaining to their lines of work.

The above definition most precisely explains Cloud computing and its usage. However, a quick glance into the quintessential characteristics of Cloud computing, Cloud computing service models, and Cloud computing deployment models will provide the readers with a better understanding on why Cloud computing plays and integral role in IoT.

3.1.1 Five Quintessential Characteristics of Cloud Computing

On-demand self-service: This refers to the facility of provisioning resources (servers, virtual machines, and databases) as and when the user requirement arises without the necessity of any human interaction or the service provider.

Broad network access: This feature refers to the uniform availability of resources over the internet to a broad variety of devices via standard access mechanisms.

Multi-tenancy and resource pooling: Multi-tenancy refers to the availability of sharing the Cloud computing resources with multiple users and customers without any disruption/intrusion to their privacy and security. Resource pooling is a feature that enables multi-tenancy to be achieved seamlessly, by dynamically reassigning the various physical and virtual resources among multiple users.

Rapid elasticity and scalability: Cloud computing provides the users with the facility of scaling up or scaling down the resource capacities depending on the dynamically changing business needs.

Measured service: The utilization of resources can be monitored and tracked, which provides a level of transparency to the users. Users and the service providers can keep an account of the activities performed with the resources and billed accordingly.

3.1.2 Cloud Computing Service Models

The Cloud computing service models are of three types.

Software as a Service (SaaS): The SaaS model allows users to use any Web-based application without any requirement to download or install the application/software on the local infrastructure. The entire computing stack is managed by the service provider. Examples include Dropbox and Microsoft Office 365.

Platform as a Service (PaaS): The PaaS model enables the customers to develop and test various applications for their business needs by providing them with a virtual runtime environment. This eliminates the need to worry about setting up the appropriate operating system (OS), middleware, and runtime environment on every individual machine. Along with the virtual runtime environment, other resources such as the storage, servers, and networking are taken care of by the service provider. Examples include Google App Engine and Amazon Web Services (AWS) Elastic BeanStalk.

Infrastructure as a Service (IaaS): The IaaS model provides the users with the entire infrastructure such as the servers, storage, and networking that is required for business operations. This eliminates the necessity for customers to worry about the infrastructure procurement, setup, and maintenance costs.

3.1.3 Cloud Computing Deployment Models

The Cloud computing deployment models are of four types:

Private Cloud: The private Cloud is an infrastructure that is hosted for use by a single organization or company. This type of hosting provides a great level of security; however, the cost of maintaining is higher.

Public Cloud: The public Cloud is an infrastructure that is hosted by a service provider and made available to customers publicly over the internet based on a shared cost model.

Community Cloud: The community model is an infrastructure that is shared among companies/organizations that fall under a particular domain such as banks and government organizations.

The infrastructure can be managed internally by the community or by a third party.

Hybrid Cloud: Hybrid Cloud is an infrastructure that is a combination of both the private and public models. This model is brings the best of the two models in terms of security, scalability, flexibility, and cost effectiveness.

After a basic discussion of the Cloud technology, we now move into the actual topic of discussion: IoT–Cloud Convergence.

3.2 OPPORTUNITIES AND CHALLENGES

3.2.1 IoT Requirements to Meet the Future Market Potential

The Internet of Things (IoT) is basically an orchestration of devices/machines that are provided with the capability of uniquely identifying and communicating with each other in order to carry out a task without the need for any human interference. The IoT in its course of development has started evolving from small standalone applications toward a vision of achieving a completely connected world. To achieve this futuristic vision, many critical aspects and requirements must be taken into account.

- With numerous heterogeneous and mobile devices being constantly connected over the internet, *ubiquitous access and uninterrupted connectivity* should be ensured for providing uninterrupted service.
- An efficient means to *dynamically collect, store, and process* the huge volumes of data that are collected or produced by numerous devices and users is required.
- The Internet of Things also has end users interacting with the edge nodes/sensors, which collect user-sensitive information pertaining to the application's usage as in the care of a health care system. Such sensitive data needs to be guaranteed privacy, security, and reliability.
- Also, each real-time application has different types of users who will look for specific services in the application. Supporting user preferences based on the context is also vital to the success of the application.
- An application can have a wide deviation in the number of users/service requests that it handles at different points of time.

Scalability of resources, enabling them to grow or shrink based on business needs, is a very import feature that ensures cost effectiveness.

A simple scenario can help the reader gain good understanding of the above-mentioned requirements. Consider a multi-specialty hospital handling thousands of inpatients and outpatients in a day. A person looking for consultation/treatment books a time slot to visit the hospital. The application basically requires consent from the user enabling the hospital authorities to access his medical records and photo ID. The app also records the user's registered phone number. This collection of user-sensitive information must be done without affecting the privacy of the user. Also, the collected information, being highly sensitive user medical records, need to be stored with utmost security measures in place.

When the user enters the premises, the CCTV at the entrance recognizes him and gives him real-time guidance and notifications on his registered mobile device. Following the instructions, the user is dynamically assigned to an available doctor for consultation. Any further tests that are recommended are uploaded in the app and the user is dynamically directed to the appropriate labs for the tests. The results are made available to the consulting doctor through the app. The doctor then analyzes the reports and specifies the medication in the app itself. The user then goes to the pharmacy and authenticates him with the registered phone number. Upon successful authentication, the medical prescription alone is made available to the pharmacist who provides him with required medicines and leaves the hospital.

The entire scenario that has been narrated from the context of a single user it has to be done for thousands of users per day. The data collected from all such users can be huge and needs a separate dedicated secure space for storage. Also, the processing of lab results and assignment of doctors to users based upon the availability must be done dynamically, which will amount to significant computational overhead. Also, the number of users who enroll for a visit will vary hugely with respect to time. During the day, we might have many users visiting the hospital and the number decreases at night. There is no point in keeping all the instances (storage, processor, and virtual machines) up and running when the demand is low. To minimize the cost and use the resources effectively, the resources should be scalable according to the needs.

3.2.2 How Cloud Comes in Handy?

The scenario is Section 3.1.1 has clearly explained the requirements of IoT to meet the future market potential. Earlier in the chapter, we also discussed the five quintessential characteristics of Cloud, its service models, and deployment models. With these in mind, it is quite clear that Cloud is the appropriate candidate that will meet all the requirements and challenges that the IoT might face in its journey toward establishing a connected future world.

In small standalone consumer applications like the smart home and care for disabled and elders, the requirement for capabilities such as processing power, storage space, and computing can be facilitated by the participating devices (mobile phones, Raspberry Pi kits) themselves. The resource requirements for such applications are low. However, for larger commercial real-time applications such as health care and transportation, the demand for resources is huge due to the enormous data that is collected and also the complexity of computations that are involved in the system. The demand for resources (storage, servers and processors) in such huge applications many a times cannot be satisfied by the participating devices alone. This is where Cloud comes in handy by catering to larger business needs.

The fundamental requirement of IoT is resource sharing. Cloud computing not only works on the principle of *sharing resources* but it also emphasizes on *maximizing the availability of resources*. One of the key features of Cloud is the *virtualization* of underlying physical devices, which makes it easier for users to share the same hardware in a seamless manner. IoT has many mobile users and devices. Services should be made available to these devices/users independent of the location. Cloud being *location independent* allows services to be accessed from anywhere and from any device over the internet. The above discussion assures that the convergence of IoT and Cloud can lead to many successful developments in the field.

3.2.3 Application Areas for IoT–Cloud Convergence

In an interconnected world with billions of IoT devices operating worldwide, data is being captured in abundance. To derive meaning of this data, it is usually processed in the Cloud and the outcome is used to improve process efficiencies, explore new revenue streams, and generate new insights into the existing business strategies. Whether it is sports analytics, health and safety at hospitals, reducing congestion and improving safety at airports, customer satisfaction in shopping malls, reducing traffic congestion in cities, or detecting component defects at manufacturing facility to save human lives, or even improve and fix artifacts in the photo captured in your mobile phone in real time, without the confluence of IoT and Cloud, nothing would be possible. It is predicted that early adaptors of Edge-enabled technologies will benefit the most.

3.2.3.1 IoT and Cloud Confluence in Agriculture

IoT devices are revolutionizing agriculture in the ways we have not imagined. Autonomous robots in a swarm multitask agricultural activities like weeding, seeding, harvesting, and hauling. These bots are modular and always in use. In addition to running on electricity, they are lighter in weight and cost less to operate than traditional agricultural equipment. These bots have multiple cameras and lidar sensors that run in parallel, which enables them to navigate in any direction. They run neural networks downloaded from Cloud and can identity rows of crops and follow them accurately. In fact, they can accurately identify weed from crop and remove them. These bots communicate with each other via wireless mesh networking and the learning can be transferred between each other. In near future, farms will be run autonomously rather than renting it out. This in turn is a huge saving for the landowner.

3.2.3.2 IoT and Cloud Confluence in Manufacturing

In today's age of manufacturing, the processes are extremely complex, precisely engineered and infinitely faster than what it was a decade ago. Every activity is timed to a second. This is mostly accomplished by automation using robotic arms and rule-based anomaly detection systems. Rule-based anomaly systems are legacy systems that lack the modern Cloud computing power and IoT.

Image-based anomaly detection systems in a high-yield manufacturing environment can process tens of millions of images per day and perform inference at multiple points per second, which cuts manufacturing time by days or even weeks on a longer run. This also helps detect more defects before entering the final assembly line. Image-based anomaly systems usually have high cameras over the edge devices where inference happens, and the training happens on an on-premises Cloud. Manufacturing industries prefer on-premises Cloud due to sheer volume of data needed to be sent online to the Cloud for processing.

3.2.3.3 IoT and Cloud Confluence in Oil and Natural Gas Rig Safety

IoT and Cloud play a vital role in ensuring safety of people working in dangerous environments. A combination of mechanical and human errors has led to loss of lives in the past and present. AI-powered computer algorithms powered by on-premises Cloud and network of cameras sensors to locate workers at unsafe locations and alert them before any accidents could happen.

3.2.3.4 IoT and Cloud Confluence in the Retail Industry

IoT and Cloud play a vital role in the retail industry. They can be gainfully employed to cut operational expenses, improve decision making, and increase revenue by billions of dollars. On an average, retailers lose 2 percent of sales due to shrinkage. For large retailers, this shrinkage would cost billions of dollars in losses. IoT devices backed up by AI in the Cloud can accurately detect mishaps like mis-scans using intelligent video analytics. These systems are proven to be 99 percent accurate and immediately inform the sales associates so that they can take appropriate action. IoT-powered store analytics tells retailers about popular store locations, unique visitors, and popular hangout areas in the store. It also lets the retailers know about the cold spots. Empowered with this information, retailers can do precision marketing and do better store merchandising.

Autonomous shopping, also known as grab-and-go stores, is gaining popularity. This kind of shopping enables customers to experience faster check-outs without a cashier by just using their mobile phones. This kind of stores reduces operational expenses and delivers a faster shopping experience.

3.2.3.5 IoT and Cloud Confluence in Supply Chain and Logistics

Supply chain and logistics is one of the most complex processes. It involves accurate and fast forecasting. Companies save millions of dollars if the supply chain gets shortened even by one day. Warehouse logistics involves cutting-edge robots with computer vision over the edge and the power of Cloud to compute millions of items by store combinations, thereby improving retail forecasting.

3.2.4 Challenges That Come with Convergence

The previous sections have justified that IoT–Cloud convergence is undoubtedly the best fit for the implementation of many real-time

IoT applications. However, there are a few challenges that might be encountered while combining the two technologies. Addressing these technologies and developing countermeasures will help in the development of effective real-time applications.

Secure communications and transfer of data: A secure data transfer between the service providers and devices/things is necessary to ensure secrecy and integrity of data. A careful choice of cryptography algorithms and appropriate protocols plays a very critical role in protecting the data in transit from being interpreted and/or tampered by eavesdroppers.

Proper identification and authentication of devices/things and Cloud service providers: Given the fact that an IoT system/application can have encounters with devices that it has never experienced before, careful mechanisms should be devised to properly identify, authenticate, and authorize the devices to ensure security against intruders. Also, the Cloud service providers being third party providers need to establish their identities and authenticate themselves to the devices/things in the instances where sensitive data is being shared for computational and storage purposes. Therefore, a two-way authentication is required in order to ensure a fool proof system.

Privacy of user's information/data needs to be ensured by using proper *access control mechanisms*, which determine the level of access that each participating entity in the system has to the data. Data is made available based on the "need to know" principle and the type of work that needs to be performed by the entity. Not all data is always required for all computations. The Cloud in most situations is also accountable for *regulating and coordinating the devices/things* in the system with proper access control policies and mechanisms.

Also, there might be certain situations where a particular entity is unavailable to perform the assigned task and the task has to be delegated to another entity dynamically. For example, in a situation where a surgeon is unable to perform a surgery due to an unforeseen situation, there must be a mechanism to *auto delegate* the task to a surgeon with a good track record. This computation should be done dynamically, and assignment should be done with no or little lapse of time.

Choice of appropriate Cloud deployment models is very critical for the success of the business. For example, an application that deals with highly sensitive data cannot rely on the public Cloud platform to store the data; however, if the application uses some high-end resources for certain computations (e.g., Graphics Processing Unit

[GPU]) that are too costly, then public Cloud comes in handy. In such a situation the organization must find a balance between the private and public Cloud (hybrid Cloud) such that it achieves cost effectiveness without any compromise in terms of security and performance.

The Cloud service provider, being the aggregator of information/data, should be accountable for the data and abide by the privacy terms and conditions. Organizations must make a careful choice of the service providers based on the service providers' reliability and efficiency in providing services.

Virtualization of resources: Advanced and specialized hardware resources such as GPUs should be effectively virtualized such that many virtual machines share one physical GPU (cost-effective). The inverse of this situation is also possible wherein multiple virtual GPUs are allocated to a single virtual machine that executes highly demanding workloads. Also, with the hugely changing demands in an IoT application, the need for resources can increase suddenly and fall sharply. Under such situations *dynamic allocation of resources* must be done in order to ensure uninterrupted services. Virtualization enables the Cloud to maintain a pool of resources and allocate the resources as per the changing demand.

A seamless transition of resources is crucial considering the ubiquitous nature of participating devices. When a device moves from one location to another, resources might have to be reallocated in order to felicitate better user experience in terms by reducing the latency involved. Also, if a particular resource (e.g., server) in a geographical location goes down, the services should be continued with fall back resources.

Considering the huge number of devices that will be connected over the internet to the Cloud (devices can also span across geographical locations), and the huge volumes of data that they generate from various sources, the Cloud should be equipped to handle the demand that might arise in terms of storage and computational power. The Cloud should be capable of *decentralized monitoring and maintenance of resources*. The Cloud should also be able to handle and organize the data coming in from multiple devices mostly in an unstructured or semistructured format. Whenever any computation or even simple storage of the data is required, it should be converted into a structured format to enable further analysis via application of algorithms and logic.

Resource-intensive computations when encountered in an IoT application can leverage the Cloud resources to complete the task and a faster and cost-effective manner. All that needs to be one is to transfer the data (*data offloading*) to the appropriate Cloud resources for computation; however in real-time applications, the complexity is increased due to heterogeneity of platforms and devices, different formats of data collected from different sources, unpredictable network conditions, and unique characteristics of the application.

The above are a few typical challenges that can be encountered in real-time when the IoT–Cloud convergence comes into the picture. Addressing these issues by incorporating effective mechanisms can help develop strong and resilient applications that can lead to the realization of a future connected world.

3.3 ARCHITECTURE FOR CONVERGENCE

3.3.1 An Overview of Cloud-based IoT Services

There are numerous Cloud service providers in the market who offer services for businesses to develop productive, scalable, and real-time IoT solutions. Most of the service providers offer specified services that help to meet the requirements and growing needs of IoT as discussed in Section 3.1.1. This section will provide a quick overview of the IoT platforms/services provided by major Cloud service providers such as Amazon, Google, and Microsoft.

3.3.1.1 Amazon Web Services (AWS) IoT Services

AWS offers numerous services that help customers in developing reliable and resilient IoT solutions in various domains. Security, scalability, data management, and analytics are the primary features offered by its services. The AWS IoT services are broadly divided into three major categories: device software, connectivity and control services, and analytics services.

Device Software
The device software enables provides the users with hassle-free experience in connecting the devices/things to the Cloud and equips the devices (also called as edge nodes) to be able to perform a number of critical computations. The two services offered under this category are:

- *FreeRTOS* is an operating system that is provided by AWS that helps in easy programming, deployment, and management of the small-sized low-end edge devices. It also eases the process of securely connecting the edge devices to the Cloud.

- *AWS IoT Greengrass* is a software that enhances the capabilities of edge devices to minimize the dependency on Cloud for all computations. It also ensures minimized latency and overhead in transporting the data over to the Cloud and vice versa by the data caching powered at the edges.

Connectivity and Control Services

The services offered in this category ensure that devices abide by specified security protocols, eases the process of registry, and connectivity of devices/things over the internet to the Cloud. The services also make it possible to manage and orchestrate the multitude of devices through the Cloud. Following are the services offered in this category.

- *AWS IoT Core* helps in easing the communication between devices and Cloud by ensuring security in the process.
- *AWS IoT Device Defender* ensures that the connected devices abide by the prescribed security guidelines by constantly monitoring the devices.
- *AWS IoT Device Management* enables proper registration, orchestration, and management of devices with the Cloud performing the job of a mediator.

3.3.1.2 Analytics Services

The growing number of heterogeneous IoT devices contributes to huge volumes of unstructured data being generated, which poses a greater challenge for the applications to process and analyze such data. AWS provides services that makes tedious jobs much easier for customers. Services provided under this category are as follows.

- *AWS IoT Analytics* is a service that enables the users to easily run complicated analytics on huge volumes of data generated/ collected by the devices.
- *AWS IoT SiteWise* service eases the process of gathering, organizing, and analyzing data that comes in at a larger scale.
- *AWS IoT Events* helps in easing the process of event detection and response that originates from the large number of sensors and applications.
- *AWS IoT Things Graph* eases the connection establishment activity between the IoT devices and the Cloud services.

3.3.1.3 Microsoft Azure IoT Products and Services

The Microsoft Azure IoT products and services enable the connection and management of numerous IoT devices. It offers a scalable and open platform for the development of Cloud-based IoT applications. The services offered by Azure are as follows.

- *Azure IoT central* enables faster connection of IoT devices to Cloud, offers centralized management of the devices, and provides application programming interfaces (APIs) and connectors that establishes direct connection between the data generated/collected by IoT devices and the decision-making services.
- *Azure IoT solution accelerators* offer predefined templates or application code for common scenarios like remote monitoring and device simulation, thereby enabling customers build customized applications using the off-the-shelf code and scale them according to their business requirements. This helps in speeding up the development process and saves effort that can be used for innovation.
- *Azure IoT Edge* helps to move certain Artificial Intelligence and analytics workloads to the edge devices, thereby providing near real-time responses to requests, reduced costs, minimized latency, and allowing offline operations or intermittent connectivity.
- *Azure IoT Hub* offers support for secure and authenticated two-way communication between the IoT devices and the Cloud (Azure).
- *Azure Digital Twins* helps to develop a visual representation or model of the physical environment with proper representations of communication between people, places, and devices. It facilitates data collection from the physical environment via models rather than relying on sensors.
- *Azure Time Series Insights* enables businesses to convert the enormous data produced over time from sensors into meaningful data. It stores time series data in near real time and provides insights from this data in minutes without any pre-preparation or efforts spent in coding. Consumers gain a global view of what is happening with their sensors and machines.
- *Azure Sphere* helps in protecting the data, infrastructure and helps maintain privacy. It provides security for the already in

place devices by means of updates and provides consumers with new IoT devices that are highly secure. It also provides certified chips with built in security measures. The Azure sphere OS provides a trustworthy platform for IoT application development.

- *Azure Maps* helps incorporate location intelligence into IoT applications. It provides facilities like search, routing, traffic insights, and provides real-time (mobility) location intelligence.
- *Azure RTOS* is a real-time operating system that eases the connectivity of embedded devices and gateways to the Cloud. It helps in developing easy-to-use applications and enhances the real-time performance of simple, low-capability devices.

3.3.1.4 Google Cloud-based IoT Services

Google Cloud IoT offers multiple tools and services that aid devices to connect to the Cloud. It likewise offers services and tools to gather, store, and process the data that is collected. Google Cloud IoT platform is completely scalable, having fully managed Cloud services. Its software stack is designed to address both edge/on-premises computing with AI infused at all levels.

Google Cloud IoT Core is a fully managed service that allows easy and secure connection and management of IoT devices. It also allows ingesting data from millions of IoT devices across the globe. Clout IoT core with other services on the Cloud provide complete solution, end-to-end retrieval, and visualization of IoT data in real time.

Google Cloud Dataflow is a service for streaming and batch analysis. It enables fast and simplified streaming data pipeline with lower latency. Dataflow automates provisioning of resources and manages them, thereby offering infinite capacity to automatically manage seasonal or spiky workloads.

Google Cloud ML Engine is used for training, deploying, and running machine learning (ML) Models. Cloud ML Engine makes it less complex for machine learning developers and data scientists to take their ML projects from deployment to production. The AI platform supports Kubeflow, which lets portable ML pipelines that can be run on-premises or on Google Cloud.

Google Cloud Pub/Sub is a service that, for Ingest connection and management, provides reliable staging location for data on its journey to processing. The Pub/Sub takes care of scaling, load isolation, and expansion of applications and pipelines to new regions. It also simplifies the development of event-driven microservices, allowing services to react to it.

Google BigQuery is a serverless, scalable Cloud data warehouse for fast querying. It can analyze petabytes of data using ANSI SQL at high speeds and no operational overheads.

3.3.2 State of the Art: Convergence Architectures

In recent times, there has been an ever-increasing demand for IoT applications. However, the real-time challenges that are being faced in realizing the IoT to its full potential have led way to a considerable research in the area. With advancements in Cloud computing, researchers have started to see the great benefits that can be obtained by bringing IoT and Cloud together. Much research has been done in developing an apt architecture that would put Cloud to a better use in IoT applications.

In reference [2], the authors have proposed a simple architecture to depict the future Internet of Things. The Cloud has been incorporated into the system to function as a storage platform to store all the information collected/generated by the devices. It also functions as a mediator/managing authority that protects stored data by implementing access control rules and policies. Also, it serves as a computational platform wherein all the huge power-consuming calculations related to access management are done. Also, the paper discusses a decentralized mechanism to help in the registry and monitoring and management of devices/things that are connected to the Cloud.

In reference [3], the authors have proposed a generalized IoT–Cloud architecture to meet the demands of IoT applications. With millions of devices being connected over the internet and demanding application-specific services, it becomes difficult to manage and serve all such requests. The IoT Cloud proposed in this paper emphasizes the virtualization of resources in order to ensure full utilization of physical machines (improved resource utilization) and scalability of servers. A virtual resource pool (virtualization technique using hypervisors) is established on numerous physical machines with all the required hardware resources. Based on the demand, the resources from the pool can be assigned to the virtual machines dynamically.

The system also includes other supporting components such as databases, application server, and load balancers to ensure availability and improved performance throughput. Additionally, HTTP and MQTT servers discussed in Chapter 2 have been used to provide services to end users and to ensure real-time communication among the connected devices, respectively.

In reference [4], the authors have presented a Cloud-based IoT architecture, which basically consists of three layers, namely, the application layer, perception layer, and the network layer. The perception layer consists of protocols and mechanisms that identifies the various objects and collects information from the environment. The network layer consists of protocols that help in smooth transmission of the collected/generated data over to the internet/Cloud. The application layer is the one which regulates the access to various services based on demand.

In reference [5], the authors have proposed an IoT–Cloud architecture that presents a model for device/things monitoring and control. The architecture is basically composed of three layers: the wireless device network, gateway, and the Cloud service. The 6LoWPAN technology discussed in the previous chapter has been used to implement wireless device network, which collects data from the various connected devices. The gateway transforms the collected data to fit the protocol of the Cloud service. The Cloud performs the job of the mediator with the help of the event manager and event rule manager services that help authoring of rules and regulating access to services.

In reference [6], the authors discuss a multi-Cloud IoT architecture called the iKaaS architecture. The architecture introduces the concept of local Cloud and global Cloud to provide an intelligent, privacy-preserving and secure big data analytics platform. The local Cloud has enough resources and capabilities (computing, storage, and network) to service requests from users/devices within a specified geographical region over a specified time period. The global Cloud serves the conventional purpose of providing dynamic and demand computation and processing capabilities. It helps to provide increased business opportunities in terms of scalability, reliability, efficiency, and, most importantly, ubiquity.

In reference [7], the authors have proposed a data-centric IoT framework. The proposed framework consists of two parts: local field and Cloud platform. The local field encompasses the personal area network, local area network, and the wide area network implemented using the Thread, WiFi, and the Long Range (LoRa) communication protocols that help in establishing the connection among the participating devices and gateways. The Cloud platform is implemented using the Azure Cloud platform services (discussed in Section 3.2.1) enabling provisioning and management of devices/things, storage, analytics, and visualization.

In reference [8], the authors have presented a novel IoT–Cloud architecture for "Big Stream" applications that have many data sources, higher rates of information exchange, and greater need for real-time processing. It aims at minimizing latency and improving the resource allocation process with the help of a listener-oriented communication model. The services or applications that are interested in specific data from a data source (edge devices) register themselves and wait upon them for any input. Based on the application's specific needs mentioned as rules, certain specific Cloud services work upon the data (normalize the data) in converting it into a format applicable/desirable to the consumer application/service.

In reference [9] the authors have proposed an edge-based IoT Cloud architecture with the aim of reducing latency and improving the security and efficiency of the system. To enhance the security aspect in the Cloud and enable IoT, the authors have adopted the traditional trust-based evaluation mechanism. A large part of the heavy resource-consuming trust calculations is done at the edge networks. This helps in reducing the latency and cost incurred in transferring the information back and forth between the Cloud and devices.

In reference [10], the authors have presented an architecture called the Internet of Things edge Cloud federation. The architecture addresses the problem faced in the transfer of data between the devices and Cloud due to the different protocol stacks and heterogeneity of devices. The model also uses the apache kafka publish/subscribe platform for fault tolerance and Kubernetes that provides a pipeline for data processing.

In reference [11], the authors have presented a first of its kind IoT architecture using osmotic computing. This computing technique involves a mechanism to manage the various resources, data, and services as the elements move across various infrastructures that are heterogeneous in nature. In order to orchestrate or manage such resource movements, the system has to be trained using an Artificial Intelligence module.

3.3.3 A Simplified Convergence Model

Figure 3.1 depicts a simplified IoT–Cloud convergence model. The architecture is built upon the most popular three-tier architecture for IoT, which includes the IoT devices layer making up the bottom or first layer. The second layer comprises one of the edge-computing technologies (mobile edge computing, fog computing, Cloudlets, etc.), which will be discussed in Section 3.3.3. The last layer consists of the Cloud layer. The edge computing layer helps to bring some of the services offered by the Cloud closer to the IoT devices layer, thereby helping to minimize cost and latency, and improve privacy, security, and efficiency of the application.

The data normalization layer lies above the edge computing middle layer, which houses several protocol connectors that act as a pipeline and helps in organizing the heterogeneous data coming in from multiple platforms into a format that is understandable to the Cloud resources, enabling it to process the data at ease and give accurate results. It acts as a bridge/mediator between the IoT device layer and the Cloud.

The Cloud apart from the various resources that it houses also has a set of supporting services that assist in effective resource allocation.

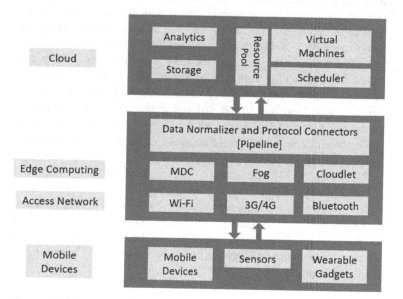

Figure 3.1   A simplified IoT–Cloud convergence architecture.

The device registry is responsible for maintaining a record or old and new devices that the Cloud encounters requests from. This enables to establish a mutual authentication between the Cloud and the device, thereby reducing the possibility for any security threats and attacks. The Cloud also has a resource pool, which houses several virtual machines that are replicas of the underlying physical machines. The resource pool has a scheduler that is responsible for dynamically allocating or de-allocating a virtual resource to a specific task.

3.4 DATA OFFLOADING AND COMPUTATION

3.4.1 Data Offloading and Computation: An IoT Perspective

The recent development in the field of IoT has led to a widespread adaptation of its applications. The IoT has several low-cost wireless sensors and devices that are connected to each other to share and process the collected data. The sensors and devices constantly monitor their surroundings and collect huge volumes of data. With the incorporation of Cloud to bridge the gap/shortcomings of IoT in terms of resources, Cloud computing has become an integral part of IoT. However, this incorporation of Cloud to IoT to offer different services means that there needs to be a huge volume of data transfer between the connected IoT devices and Cloud. Although offloading of data from devices to Cloud helps overcome the limitations of IoT, high additional cost is incurred and maintenance is required for communication links.

The Cloud in most cases being located far away from the users requires large amounts of data to be transferred to and from the Cloud. This movement of data causes several performance issues like heavy loading of servers, longer response times, lowered bandwidth for communications, and increased costs. All these limitations have led to an increasing demand in finding alternative data offloading techniques. One of the common techniques includes setting up *a terminal to terminal (T2T) network* that totally depends upon the direct communication between mobile nodes. This communication therefore does not require a backbone infrastructure for communication. This is more suitable for applications that gather and use data that is delay tolerant. Such *time-insensitive data* can be withheld in a particular place before it can be transmitted over to a distant Cloud service or infrastructure for analysis.

Another alternative is the use of *mobile data offloading technique* wherein the network has many mobile nodes. In situations where the network is heavily loaded with traffic or is not accessible due to the geographical location, the network provider may decide to send the delay-tolerant data to only a few mobile nodes within the specified range. Later, when these mobile nodes meet other mobile nodes (that are interested in the content/data), they exchange the data by means of peer-to-peer communication (P2P). A variant of this technique is the prediction-based offloading mechanism, which functions based on the *predictable behavior of human mobility*. Studies have shown that the mobility of human beings follows a particular pattern and has a correlation to the locations forming clusters [12]. Based on this theory, data is offloaded to mobile nodes that are more likely to transfer the data to the required destination whenever the opportunity arises (opportunistic peer-to-peer data exchange).

Although the above discussed mechanisms are suitable in most cases, certain applications cannot afford the delay caused by adopting them. Applications that collect and process time-sensitive data consequently work on strategies that improve the efficiency by offloading the data, processing, and storage closer to the sources that collect information/data by adopting the edge computing technologies.

3.4.2 Edge Computing Technologies for an IoT Network Infrastructure

Edge computing is a mechanism used to enhance the data processing, computing, and storage activities in applications/systems where the Cloud is an integral part. The processing (computing and storage) capabilities are brought closer to the networks' edge (data sources), thereby leading to several advantages.

- It helps in reducing the latency involved in transporting the data to and from the Cloud.
- It also helps to perform computations offline without dependency on the Cloud for all processing activities.
- The performance and throughput-related challenges that occur due to the intermittent internet connections in most cases can also be overcome.
- Reduces the load on Cloud servers and helps reducing the load on communication links (congestion) due to the transport of huge volumes of data (bandwidth-intensive big data).

- It helps in near real-time data processing and analytics closer to the data sources.
- It reduces the organizations' operating costs and the scheduled downtimes (from the Cloud providers).

Following are a few edge computing technologies. All the below discussed technologies form the middle tier of the three architecture, that is, the IoT devices, edge/fog/Cloudlets etc., and the Cloud.

Multi-Access Edge Computing or Mobile Edge Computing (MEC) is an architecture that is implemented to lower the network congestion for improved performance of the application. It also satisfies the applications' needs such as higher bandwidth, lower latency, and location awareness. The architecture enables the resources that make up the Cloud to be placed within the radio access network (RAN). The MEC in most cases are developed using open source hardware and software like SDN (software defined network) and NFV (network function virtualization) that make use of the facilities provided by Cloud and adopts the virtualization mechanisms of the underlying hardware on which it runs. The virtualization infrastructure is housed in the MEC's mobile edge host (MEH) layer. The MEH is the key component of MEC that functions as the application server enabling the movement of storage, computing, and processing capabilities closer to the edge of the network.

Fog computing is an architecture that is implemented with distributed data centers closer to the edge devices. Individual service providers build distributed data centers or distributed Cloud, which houses various Cloud computing resources. The resources that offer services can be placed in a central office or within the customer premises also. It is more like MEC architecture in bringing the Cloud capabilities closer to the network's edge. However, in fog computing, a common centralized computing server processes all the data from various endpoints in the network, while in edge computing, every network performs data processing. Another major difference lies in the positioning of the intelligence and computation capabilities. Edge computing moves all or most of the processing and computational capabilities into the edge devices, while in fog computing the processing capabilities and intelligence are pushed into a local area network (LAN) that houses various fog nodes. The fog nodes receive the near real-time data and perform analysis on them (for time-sensitive data) or forward the data to the Cloud for further action (for time-insensitive data) depending on the use case and nature of data received. Fog computing uses local data (that is already present in the fog nodes) and the data received from the IoT devices and perform near real-time processing of data. Unlike MEC, since the fog computing depends upon a single centralized server, there is a huge possibility of service disruption to occur.

Cloudlets are a group of computing resources/data centers implemented on a smaller scale that are hosted in order to improve the computing capabilities of mobile devices within a geographical location/proximity. It basically helps to overcome the latency issues that are encountered by mobile devices in a traditional wide area network (WAN) by hosting the Cloud resources physically closer to the mobile devices that require them. A Cloudlet is particularly different from a public data center in that the Cloudlet is personally managed by the users/businesses that use it while the public Cloud is managed solely by the Cloud service provider. As the Cloudlet is locally maintained, it obviously depends upon the LAN for connectivity. Cloudlets are thus used by a smaller, specific, and closed group of users unlike the public Cloud. It is basically used in resource-intensive applications that involve speech recognition, virtual reality, machine learning, augmented reality, and language processing.

Micro Data Centers are small-scale customized data centers with all the required Cloud resources in terms of computing, processing power, and storage that can be deployed closer to the edge devices and in locations where a traditional data center cannot be practically set up. It thereby reduces cost and latency issues. While the fog computing is a form of distributed architecture, micro data centers (MDCs) are a type of data center design. MDCs typically help to build the edge computer in a real-world scenario. Businesses can scale up progressively based on their changing demands. MDCs facilitate easy deployment as the required equipment are preinstalled by vendors on server stacks and shipped as is. Upon receipt all that must be done is just an installation of any additional local components followed by connection to an existing power supply. Once the MDCs are up and running, service providers configure and monitor the equipment using various tools.

Cloud of Things is a mechanism that makes use of the edge devices' resources such as advanced-feature mobile phones and other handheld devices to form a virtualized Cloud infrastructure. Here the IoT devices themselves communicate with each other and perform computations and processing of real-time data without depending upon the Cloud. For example, a car can send traffic-related updates and warnings to other cars and suggest alternate routes without the need for a user to feed in data.

3.4.3 Data Offloading and Computations

With the recent increase in usage of IoT applications that collaborates with the Cloud for many resource-intensive computations, of huge volumes of data collected at the edges are transferred to remote sites for processing. Although offloading is important in order to perform computations and storage at a larger scale, it incurs significant cost, causes latency, and lowers the bandwidth due to the transportation of data to and from the Cloud. Is also poses security and privacy threats to the data in transit. A considerable amount of research is being done in this area in order to suggest efficient and optimal offloading techniques to improve the system performance and reliability.

In reference [13], the authors have proposed prediction-based offloading schemes that predicts any available opportunities for transferring data in future between mobile nodes in a terminal to terminal (T2T) network based on the topology varying with time, node mobility pattern, and the temporal contacts. The proposed system has been verified to provide results with improved delivery ratio and reduced latency overhead in comparison to other methods.

In reference [14], the authors have proposed a genetic algorithm based adaptive offloading (GS-OA) for efficient communication in the IoT–Cloud Infrastructure. The mechanism serves to improve the response rate to requests with the help of the fitness process that is distributed among the gateways and other devices in the infrastructure. This in turn leads to lesser delay and processing time in the system, leading to better performance and improved throughput.

In reference [15], the authors have proposed a fog-based data offloading mechanism to provide a reliable means of data collection and improved scalability in dense and dynamic environments. The systems construct a multi-tier data offloading protocol with the help of the fog networking concept, thereby supporting a variety of data-centric IoT applications in the urban environments. The sensors present in the heterogeneous network offload the data among themselves or to the mobile gateways. The data drop-off rates are monitored and minimized based on the data pertaining to the mobility of humans and the urban environment in which they operate upon. As a result, this mechanism provides a lightweight and efficient method for data offloading.

In reference [16], the authors have attempted to solve the traffic congestion problems encountered in IoT applications by proposing a game-based data offloading mechanism that leverages the Vickrey–Clarke–Groves (VCG) mechanism and Rubinstein bargaining model. The proposed scheme also addresses another aspect of data offloading, which is incentivizing the access point owners. Here the mobile network owners are responsible for properly incentivizing the access point owners so that they will offload the traffic of non-registered users to the network as intended.

In reference [17], the authors have proposed an efficient mechanism that saves time and energy in the process of selection of an appropriate fog device (FD) for offloading. A module placement method by classification and regression tree algorithm is presented wherein decision on the selection of the best FD is done based on the power consumption of the fog device and other parameters such as authentication, confidentiality, speed, availability, integrity, capacity, and cost.

In reference [18], the authors have presented a computational offloading mechanism to address the challenges encountered in offloading heterogeneous and huge volumes of mobile data to arbitrary resources. The mechanism in turn claims to improve the efficiency, time, and energy consumption of mobile devices. In the process, the system adopts the non-dominated sorting genetic algorithm III (NSGA-III) to solve the problem of multi-objective optimization in task offloading in edge computing.

In reference [19], the authors introduce a framework that enables device-to-device edge computing and network (D2D-ECN) for data processing and computational offloading. This includes many resource-rich devices that participate in the network for offloading and computation. However, such devices are prone to fall short of battery life when involved in such high-efficiency computational tasks. Hence an energy harvesting technology based on reinforcement learning is incorporated to the D2D-ECN framework. Also, an online approach with the underlying principle of Lyapunov optimization is used for efficient offloading and resource management in dynamic environments.

In reference [20], the authors have proposed a nested game-based computational offloading for mobile Cloud IoT (MCIoT) systems. The MCIoT extends the capabilities of mobile and portable devices, enabling them to execute different mobile applications. The Rubinstein approach is employed to determine the remote offloading computation part and then a Cloud resource is dynamically assigned for the computation. An optimal solution for this offloading process is achieved by employing the nested game approach.

3.4.4 Offloading Considerations and Challenges

Offloading basically is a mechanism that helps in moving the resource-intensive and energy-consuming task to high-end resourceful devices, thereby reducing delay, increasing bandwidth for relevant communications, and energy conservation. Offloading can also be done to store data that may not be required in near future (archival data). However, to obtain an optimal throughput and performance of applications there are certain *considerations* with respect to implementing offloading in the IoT systems.

- As the process of offloading by itself incurs cost, offloading cannot be done all the time depending upon only the users' requirements. A decision on *when* (based on the current situation) the offloading can be done is very crucial. For example, consider the activity of load balancing among multiple fog nodes. Offloading the data does not help or is inefficient if all the nodes are minimally loaded or extremely loaded (i.e., equal proportion). Offloading helps only if the load in multiple servers/nodes differ greatly (i.e., one in heavily loaded and the other is sparingly loaded).
- Not all data/computations need to be offloaded. Certain computations that can be accomplished with minimal resources need not be transferred all the way to the servers, which will only add additional overhead. *What* needs to be offloaded is an important decision to be made.

- Depending upon the context and type of computation to be performed, a decision on *where* the data is to be offloaded must be made carefully. Despite the Cloud providing all the required services in an efficient way, the latency involved in transfer of data makes it debatable. Hence, alternate solutions discussed in Section 3.3.3 can be taken into consideration.
- A decision on *how* to offload is also very important in the IoT systems. We can offload the tasks at different levels depending upon the applications needs as follows:
  i. *System level offloading* is a mechanism wherein the entire the complete edge application along with the operating system on which it was running is pushed to the server/ Cloud (e.g., CloneCloud).
  ii. *Application-level offloading* is a mechanism wherein the multiple applications are run as services on servers and the user/client can use these applications in their system by accessing them via Web services (e.g., DAvinCi)
  iii. *Method-level offloading* is a mechanism where the application/program is broken down to separate modules/pieces and depending upon the need certain modules are run remotely on servers (e.g., remote procedure calls (RPC)).

Following are a few challenges encountered in the process of offloading.

Scalability and resource allocation: Allocation of the desirable number of resources at the location of task execution (i.e., edge or Cloud) in order to improve the efficiency is a challenge. There may be extreme situations where resources are either idle/under-utilized due to lesser load or resources heavily loaded demanding frequent offloading of workload. Under such circumstances, scaling up and down of resources respectively for effecting utilization remains a challenge.

High software quality: Offloading framework partitioning and users' varied inputs affect the system performance greatly. High-quality software is required to optimize the framework portioning and improve user experience.

Scalability of applications: Scaling up the applications to service requests coming in from multiple users without affecting the offloading process is a challenge in maintaining the performance.

Service level agreements (SLA): When critical IoT applications that handle sensitive data perform offloading, care should be taken that SLAs are not breached. Trust on the service provider and SLA monitoring are critical factors to be followed up during offloading process.

Security and privacy of user data: In the process of offloading, the threat of data being interpreted, tampered, and misused increases when in transit. Ensuring a secure transmission medium and a resilient framework for the security and privacy of user data is of prime importance.

Integration and interoperability: With the confluence of a variety of heterogeneous networks, platforms, operating systems, and devices in the IoT, integrating all of them to perform tasks in a harmonious manner is a great challenge. Different service providers and IoT middleware might employ different protocols and privacy and security mechanisms, making it difficult to synchronize them and bring about an organized behavior in applications.

Energy consumption: The offloading process by itself consumes energy and bandwidth, thereby reducing system performance. An informed decision must be made about whether to go ahead with the offloading process or not.

3.5 DYNAMIC RESOURCE PROVISIONING

3.5.1 The Resource Provisioning Activity and Requirements for IoT

Resource provisioning is a mechanism by which Cloud services including storage, infrastructure, network, and computing capabilities are made available to the applications/users on a pay-per-use basis via various Cloud computing vendors. The actors (software agents, real-world objects, and human beings) in the IoT architecture collaborate with each other in real time and share information, which helps individual networks to connect to the backbone, thereby helping the IoT to be realized over a wide range of objects. The IoT sensors are capable of monitoring and collecting information from the dynamically changing environments. To enable efficient functioning and processing of this data, effective resource allocation/provisioning is of prime importance in order to carry out the necessary tasks. One of the major challenges in resource provisioning is the allocation of an apt amount of resources for performing the task while still minimizing the cost and maximizing resource utilization. Resource provisioning is categorized into three types, namely, static resource provisioning, dynamic resource provisioning, and user self-provisioning.

Static resource provisioning refers to a mechanism wherein the resources are allocated in advance by means of a contract between the customer and service provider for the necessary services. The service provider prepares and provides the resources beforehand and charges the customers a fixed/flat fee or on a monthly basis. This is generally used for applications where the workload is predictable and the demand for services remains unchanged for a long period of time.

Dynamic resource provisioning is a mechanism wherein the resources are allocated by the provider based on the changing needs of applications. Here the virtual machine (VM) technology is used, which helps underlying physical resource to be separated into multiple virtual devices, thereby facilitating multiple tasks to execute simultaneously and reallocation to support the dynamic demands. All such virtual resources are maintained in a pool and based on the incoming workloads to the Cloud, the scheduler makes an informed decision about the allocation and provisioning of VMs for task execution. The scheduler is run at periodic intervals to perform the following activities:

- Predict the probable amount of inflow of workloads in future.
- Determine the amount of VMs to be provisioned from the Cloud.
- Monitoring the waiting time of jobs in the queue and deciding on allocation of VMs to the deserving jobs.
- Monitoring the VMs and release of idle VMs.

In this case, the users/customers are charged on a pay-per-use basis. *Cloud bursting* is a term used when a hybrid Cloud is created using the concept of dynamic provisioning.

User self-provisioning is a mechanism wherein resources are available for purchase to the users via a Web form where they create an account and pay for the resources using a credit card. The resources are then made available with a few hours.

When considering an IoT scenario, the best fit resource provisioning mechanism is the dynamic resource provisioning considering the dynamically changing environment and needs of the application.

3.5.1.1 Resource Provisioning Requirements for an IoT-based Environment

The key features to be considered in the process of dynamic resource provisioning for IoT environments are categorized as follows:

- *Cost-aware* allocation refers to the overall price in terms of revenue, resources, and profit. Maximizing the revenue and profit and the cost-effective allocation of resources to users (minimum expense) is important in an IoT environment.
- *Efficiency-aware* allocation refers to the attempts made toward improving the performance (number of tasks that are run to completion) of the system by reducing the time taken to respond to and execute a task/workload. Also, an increase in the execution speed (the amount of time the workload spends in running to completion on an allocated resource) and bandwidth (the amount of data that a particular connection can handle over a specified period) with importance given to priority of tasks (based on waiting time or type of workload) is important for an efficiency-aware system.
- *Load-balancing-aware* allocation refers to the ability to distribute the Cloud resources among multiple users. It also refers to the capability to efficiently distribute the workload among different resources across data centers without any disruption to task execution.
- *QoS-aware* allocation refers to the process of improving user experience with respect to availability, throughput (total number of tasks executed in a particular period), response time (time taken to service/ attend to the request once it is out of the waiting queue), security, and reliability (assurance to lowered or no failure and guaranteed execution of task).
- *Power-aware* refers to the ability to lower the level of energy consumption across all levels in the IoT architecture. Energy is basically required for resource preparation and execution of the task in the IoT architecture.
- *Context-aware* allocation means the ability system to collect information from its environment dynamically and change its behavior in a guided manner to enhance performance and user experience.
- *SLA-based* allocation means the ability of the system to ensure that the SLAs (agreement between the service provider and

users) are not violated, thereby providing fulfilled and satisfactory services to the user.

- *Utilization-aware* allocation refers to the ability to improve the efficiency of the system by improving the utilization of resources with reduced or no idle time, thereby reducing costs incurred for the users.

3.5.2 Dynamic Resource Provisioning

In this section, we shall be exploring a brief overview of the existing approaches employed in realizing dynamic resource provisioning. In Ref. [21], the authors have proposed a priority-based pre-emption policy that helps in addressing the problems of resource allocation in a virtualized environment. As discussed, earlier virtualization helps in dynamically allocating resources in order to ensure timely execution of tasks and improved efficiency in the system. The underlying principle of the proposed methodology is based upon the priority of the job, which is decided based upon the deadline of the job. A job with higher deadline (lower priority) should not postpone the execution of a job with lower deadline (high priority). Allocation of VMs and thereby the services before the deadline is the focus.

In Ref. [22] the authors have presented a solution based on fog computing that provides a context-aware framework that has decision-making modules dispersed across the IoT Cloud platform and gateways. The solution presented makes decisions by analyzing the data input from various sensors and the context of the surrounding environment obtained from the fog computing resources. It also manages the service provisioning activity based on the changes in topology. Both the activities of service provisioning and decision making are accomplished in one framework. Experimental results have proven that the framework is resilient and responsive to changes in topology.

In Ref. [23], the authors have proposed a mechanism that provides a latency-aware distributed resource provisioning solution for edge network-based IoT applications. An algorithm that supports the distributed allocation of resources in order to support flawless integration and deployment of various applications in the IoT infrastructure is also presented. The algorithm performs a mapping of the application to the edge network/device and supports dynamic migration of the application fragments abiding by the SLA.

In Ref. [24], the authors introduce a QoS provisioning framework (QoPF), which is dynamic in nature and applicable to service-oriented IoT. The framework uses the backtracking search optimization algorithm (BSOA). The proposed framework maximizes the IoT application layer's composite service quality by balancing the service's reliability and cost incurred for the computation time.

In Ref. [25], the authors have addressed the issues related to user's privacy that arises during the collection and transport of user data in the IoT environment. The European Union has made it illegal for the IoT devices and other sensors to collect people's data without obtaining consent from the user. To ensure such user privacy, a data filtering mechanism has been introduced at the edge devices/network. Application of privacy rules to the data at the edge of the network will enable restricted transport of data to the Cloud. However, this additional processing can introduce some unexpected delay and bottlenecks in the system. Hence as a countermeasure the authors have introduced a hybrid data processing technique to address the issues of privacy and resource provisioning.

Fog computing being more suitable to address the recent and upcoming demands in IoT applications has become an integral part of the IoT architecture. With this in mind, the authors in reference [26] have presented a container-based virtualized networked architecture. The architecture finds its place in the middleware layer and works by taking advantage of the native Container Engines, thereby enabling the real-time up scaling and down scaling of the virtual resources such as computing and networking resources. The virtualized computing and networking resources are dynamically managed by a less complex penalty-aware-bin packing heuristic.

In reference [27], the authors have introduced a QoS-aware framework for dynamic resource provisioning in fog computing. The framework addresses the dynamic deployment and release of application services in fog nodes, thereby lowering the cost and latency while still maintaining the QoS requirements. Two highly efficient greedy algorithms have been used and the framework proves to be efficient even with very minimal IoT node information and makes no assumptions on the network.

In reference [28], the authors have proposed a model-based service provisioning mechanism that is multidimensional and multilevel in nature. The platform renders the potential of large-scale heterogeneous resources as lightweight services. An integrated message space has been introduced to aid the sharing of sensory information/data in a distributed IoT environment. The platform also encourages sharing and reuse of resources.

3.6 SECURITY ASPECTS IN IOT CLOUD CONVERGENCE

Form the above discussions, it is evident that Cloud computing has indeed evolved with new features in order to provide for the needs of IoT. These features include advanced processing and decision-making capabilities, virtualization mechanisms, and analytical tools. Most of these features have been discussed in the previous sections. However, all these advanced capabilities come with various risks and threats to the security of the IoT system. This chapter aims at discussing the various security challenges that accompany IoT–Cloud convergence.

3.6.1 ENISA's Categorized Security Challenges in IoT Cloud Convergence

The Cloud being an integral part of the IoT has many security-related issues due to the numerous devices with diverse specifications and protocols that are connected to it. The European Union Network and Information Security Agency (ENISA) categorizes the security aspects that come with the IoT Cloud convergence based upon the IoT reference model as follows:

- *Connectivity* refers to the interfaces (communications and interactions) among the components such as the endpoints (devices, sensors, actuators etc.,), gateways and the Cloud. According to ENISA, the two major contributors toward the connectivity-related security issues are the *heterogeneous protocols that are used for communication* and *the insecure flow of data from the edge devices into the IoT system*. The diverse range of applications where IoT is used requires the incorporation of multiple heterogenous hardware, protocols, operating systems, network connection, and so on in the IoT system. In certain applications, there is a need to integrate legacy systems with new devices, which in turn could lead to unprecedented risks in the security of the system.
    Another aspect of the IoT system includes the processing of data, which in an IoT system is done at the Cloud or at the

edge devices as discussed in Section 3.3.2. Edge computing enhances the processing capabilities of edge devices, thereby providing major advantages such as reducing the latency involved in transfer of data to the Cloud and back to the devices, reducing costs, bringing near real-time operations closer to the data sources, and many more. However, the restricted capacities of the edge devices in terms of storage and computing power make it difficult to incorporate robust security mechanisms and algorithms, thereby making them vulnerable to security threats. Faulty implementations can lead to a compromise on the integrity, confidentiality, and the privacy of the data.

- *Analysis* includes data processing, filtering, and aggregation that comes into the various levels of the IoT system, which is collected by the endpoint devices. ENISA has identified that the *real-time processing of data at the edge devices* and the *decentralization of Cloud* are two aspects that majorly contribute toward compromising the security of the IoT system. The very nature of the environment in which the edge devices operate (common places where security implementations are nil) makes it easier for attackers to gain access and manipulate the data. Also, security functionalities such as monitoring and tracking which is performed at the Cloud becomes difficult to implement and manage at the edge. Additionally, Cloud decentralization that enables better management and processing, thereby reducing the overhead on a single remote Cloud, also comes with its challenges. The edge devices that lack elasticity unlike the Cloud are more prone to denial of service (DoS) attack following which the coordination of security mechanisms such as security patches and software also increases.

- *Integration* represents Cloud APIs and other such features that facilitates the real-time bidirectional data flow in the system. According to ENISA, security can be greatly influenced by the kind of *environment/vertical that the Cloud serves*. For example, an IoT system that functions at a critical operations environment such as a smart hospital, security is greatly influenced based on the complicated infrastructure and the management policies. Security also depends upon the *knowledge and implementation expertise levels of the developers*. Developers should be completely aware of the primary elements of security in the smart devices such as message authentication and encryption, authentication and identification of things, message integrity, secure booting, key management, and necessary storage implementations. Also, it becomes difficult to lay out a generalized and standard hardware/software development life cycle for the IoT systems due to the involvement of varied degree of hardware, protocols, and network connectivity mechanisms.

3.6.2 IoT Security Designs Based on Edge Computing

Edge computing has been discussed previously in many sections. Having touched upon the basics of edge computing and the advantages it has to offer to the IoT system, it is evident that edge computing offers a great deal in optimizing the system performance. However, the devices are also vulnerable to security threats like any other system and hence arises the need to implement security solutions on these edge devices. Although edge device security implementations are at its infancy, considerable amount of research has been conducted in the area. This section aims at discussing few of the notable security solutions for edge devices.

A quick look at the edge-centric IoT architecture will provide a clear understanding to the readers on the design and deployment of the security solutions in the appropriate layers. Figure 3.2 represents

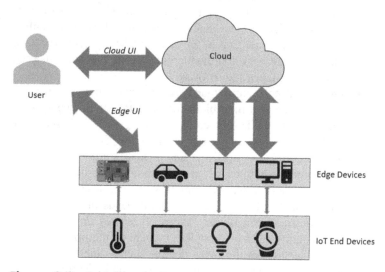

Figure 3.2 A high-level edge-centric IoT architecture.

a high-level edge-centric IoT architecture. At the lowest level are the IoT end devices (thermostat, electric bulb, smart watches, displays etc.,) that act as sensors and collect information from the surrounding environment. The next layer comprises the edge devices that are equipped to store and process considerable amount of data (Raspberry Pi, autonomous cars, smartphones, personal computers etc.). These edge devices assist in performing computations and other activities for real-time scenarios. The final layer is the Cloud wherein resource-intensive operations, decision-making, and more are carried out.

The edge layer also acts as the best candidate to deploy/implement the security solutions for various reasons. First, as mentioned earlier the edge devices being equipped with better storage and processing capabilities in comparison to the end devices, thereby making it feasible to implement complex security operations such as encryptions, attribute-based access control schemes, and so on. Second, real-time security needs can be catered to by the edge devices due to the proximity to the end devices. Third, as mentioned earlier, the edge device collects and stores the information collected by end devices for further processing. This makes it a suitable place to support security-related decisions as the collected data can supply the required information. For instance, more data helps in detecting intrusions more efficiently. Fourth, considering the widespread presence of end devices, setting up firewalls across the end devices is very intimidating. However, firewalls can be deployed at the edge layer, which in turn can filter any malicious requests/attacks. The highly mobile end devices can be monitored and tracked by the edge devices efficiently. Also, as the end devices constantly send information to the appropriate edge devices, trust can be effectively established between the edge and end devices. This eliminates the overhead of establishing trust between the end devices themselves. Lastly, the edge devices are most often connected to the Cloud via high-speed links, which makes it easier for edge devices to rely on the Cloud for any resource-intensive security computations and support. With the edge-centric IoT architecture in mind, we move on to discussing a few and most relevant edge-based security implementations for IoT.

User-centric edge-based IoT security architecture: The end users of the IoT applications play a major role in making the IoT applications operational. They gain access to many IoT resources via terminals such as the smartphones, personal computers, smart watches, and so on. The resources are made available to the users in a more convenient and pervasive manner. However, users who access these resources cannot be expected to access the resources via secure and trusted terminals. Also, not all users possess the knowledge and awareness of various security measures. Hence, management of security must be incorporated at a more reliable layer, which is the edge layer. This includes setting up measures to manage personal security of users and positioning virtual security systems at the network edge.

Device-centric edge-based IoT security design: Devices in the IoT architecture collect information from the surrounding environment and act as actuators that help in driving the decision-making process that control the physical environment. In the device-centric edge-based IoT security design, the security controls are offloaded from the end devices to the edge layer. It provides customized security solutions for each individual device based on data that the device is sensing, the effect of actuating a task as a response, the availability of resources, and the security requirements of a collection of end devices. EdgeSec [29] and ReSIoT [30] are a few designs that represent the device-centric security architecture.

End-to-end IoT security: A security system that covers all the components of the IoT system and the communications between them is ideal for realizing a reliable application. However, the heterogeneity of the participating devices, Cloud, and the protocols make it difficult to attain the goal. Researchers have identified the edge layer as a suitable candidate to house the security controls as it operates as an intermediary between the heterogenous devices and the Cloud. A secure middleware setup at the edge layer helps in establishing a secure end-to-end communication among the participating devices [31]. The middleware manages security functions such as the encryption algorithms, authentication mechanisms, MAC algorithms, and secure sessions for devices.

Firewalls implementation at the edge layer: Heavy resource demanding applications like firewalls cannot be run on resource-constrained devices. Also, the huge number of IoT end devices that are connected ubiquitously to the IoT makes it extremely costly to deploy and manage the firewall on each individual device. Hence, installing firewalls in the edge layer is an ideal choice due to the following reasons.

- Firewall updates can be more achievable due to the existence of one centralized conceptual firewall.
- As in most cases where an edge device manages a subsystem of the IoT application, the firewall can be deployed in the edge device that manages the entire subsystem.

- User mobility can be traced with the edge device monitoring the user movement and credentials of the users and devices.

Firewall policies that are defined by the IoT application are translated into flow policies. These flow policies are revisited to rule out any conflicts after which the policies become a set of firewall rules that are installed at the edge. All incoming and outgoing traffic are then scrutinized against these set of rules. Examples of firewall-based security at the edge layer include FLOWGAURD [32] and a distributed firewall architecture [33], which use software-defined networks (SDN) and virtual network function (VNF) respectively.

Intrusion detection system at the edge layer: In 2016, a successful distributed denial of service (DDoS) attack was launched by attackers by compromising a great number of IoT end devices. This disrupted internet services in a wide range of areas. The presence of an efficient intrusion detection system would have helped address the issue, thereby minimizing the loss incurred. Designing an intrusion detection system at the edge layer would be more efficient considering the availability of huge data/information that is transferred from multiple end devices/sources. This data can then be run against machine learning algorithms to identify any intrusions and adapt to changes in attack patterns. A typical design of the intrusion detection at the edge layer as shown in reference [34] consists of a distributed traffic monitoring system that collects the network traffic in real time. The data is then run against the intrusion detection algorithm locally at the edge device. The collective data from multiple edge devices are also collaboratively monitored for any intrusion. Finally, the network controllers installed at the edge devices impose the detection outcome for further action.

Edge-based authentication and authorization: Authentication and authorization play a major role in security that helps prevent many attacks such as unauthorized access and the DDoS attacks. However, the huge number of devices and heterogenous nature of platforms and protocols make it difficult to establish a strong security mechanism. Also, standard security aspects such as the digital signatures cannot be applied to the IoT ecosystem. These problems can be resolved by the edge layer that acts as a middleman between the end devices and the Cloud. The authentication process can be broken down into multiple segments with the edge being the pivot. The first segment involves the authentication between the end device and the edge, and the second segment includes the authentication between the edge and the Cloud/user/other end devices. Authentication protocols can be tailored depending upon the characteristics of the participating entities in the segment. The edge also acts as a proxy for the end devices in the authentication and authorization process. With the edge devices having the necessary resources to support the computations and processing power required for the algorithms the end-to-end authentication and authorization is made possible in the IoT ecosystem.

Edge-based privacy-preserving techniques: With numerous users connected over the internet and in turn the IoT applications, they share information that is required for basic communication and carrying out task in the system. Despite the IoT applications being an inevitable part of everyone's lives, users require strict privacy-preserving mechanisms in place. With edge layer being rich in information/data collected by the end devices, it is ideal to implement various privacy-preserving algorithms at this layer. When queries are presented via an IoT application, the edge layer processes the information against the defined goals and algorithms in order to provide the applications with data that is privacy preserved.

3.7 IOT–CLOUD CONVERGENCE: TEST BEDS AND TECHNOLOGIES

This section introduces us to a few testbeds that is a part of the FIESTA-IoT project [35]. The project aids in experimenting new approaches toward the development and deployment of IoT applications by leveraging the possibility of interconnection of varied test beds and platforms. It also aims at making the process of sharing data between various testbeds and easy and seamless task to enable research and development.

3.7.1 Overview of Testbeds and Platforms

3.7.1.1 SmartSantander Testbed

The SmartSantander is an experimental testbed. It is used for experimenting various architectures, future technologies, and services targeted for Internet of Things and applications. It is usually done in the context of a city. SmartSantander is platform scalable with new and diverse variety of sensors. The architecture of SmartSantander testbed is divided into three tiers.

IoT tier: This tier is responsible for sensing the data of interest like temperature, chemical composition of atmosphere, sound, light, presence of a vehicle parked, temperature, and humidity of soil. Sensors are built with networking capability. The network module built in them is called repeaters. Some of the sensors are standalone, in that case they use wireless technology to communicate with the repeaters. Few of these devices are powered by batteries due to the location in which they are placed. Repeaters are located at high-rise places like streetlights, public buildings, and relay the information captured via sensors. Digimesh also known as IEEE 802.15.4 is the preferred mode of communication between repeaters and IoT nodes. For experimentation, a native Digimesh interface is available in the repeaters.

Gateway tier: Gateway receives all information that is sent from IoT nodes and repeaters. The received information is forwarded to the top layers of SmarSantander intelligently and efficiently. It additionally has tons of networking tools and control and manage the network.

Platform tier: This topmost layer can ingest huge volume of data from the gateway. It has various interfaces to handle the incoming data. Platform management tools are also available here. IoT device registration repository is available in this layer. Information received from the sensors is also stored in this layer.

3.7.1.2 UNIS Testbed

UNIS Testbed has three tiers:

Server tier: The server tier is the entry point for users to access the test bed and perform operations on it. All the back-end components and its functionalities are hosted in the server tier.

Gateway tier: The gateway tier is the backbone of the test infra-structure. It connects the server tier and the IoT tier via WiFi or Ethernet. Gateway tier is also known as an embedded gateway (GW).

IoT tier: As the name suggests, IoT tier consists of IoT sensor nodes. It also comprises consumer electronics such as smartphones and smart displays. IoT nodes can be connected in two ways. The first way is wireless communication that includes WiFi and Bluetooth. Second, a USB connection, which is wired, is used for management of gateway.

3.7.1.3 KETI Testbed

The KETI testbed built for indoors. It is an experimental test bed. All indoor parameters such as human presence, humidity, indoor temperature, and indoor luminance is monitored. Armed with this data, an energy management system is built for the maintaining optimum operational level of the building. KETI testbed is divided into three tiers. They are server platform known as Mobius, device platform known as &Cube, and final IoT device tier.

Mobius (server platform): The Mobius platform complies with oneM2M standards. It is an open IoT service platform. A large number of IoT devices are indexed so that devices of interest can be easily discovered. RESTFul APIs are the preferred way of data transfer and it complies with oneM2M standards.

&Cube (the device and gateway platform): It is designed on JRE (Java Runtime Environment) to serve as the oneM2M standards-defined software platform for its home gateway and Cloud platform. It could be ported anywhere to gateways or IoT devices that can run a Java Virtual Machine (JVM).

IoT devices: IoT devices collect indoor sensory data. They transmit the data to the gateways using Zigbee wireless mesh network. Gateways are Raspberry Pis and installed with &Cube, which is a software-based gateway. JSON or XML is used to transfer the data to the server platform Mobius.

3.7.1.4 Com4Innov Testbed

Com4Innov is an actual life-sized laboratory. Its environment is already set up and in use. Com4Innov test bed is used to analyze wireless technology and its corresponding services before the technology could be ready for the market. Few examples are network applications and mobile services of the future, next gen communication objects, and smart devices for IoT are accompanied by data handling and compute facilities.

FIESTA-IoT project drives Com4Innov via 5G communication technology. Huge amounts of data are generated due to 4G/LTE network interconnection with IoT infrastructure. Com4Innov data center can host several types of services, for example Mobius from KETI testbed and OpenIoT.

Networking technologies: Com4Innov has a 4G/LTE core and an access network. A highly noncentralized architecture of sensors with methodologies and tools for monitoring and measuring is used in the Com4Innov test bed.

Applications and services: The applications that are available in Com4Innov test bed are IP Multimedia Core Network Subsystem and Applications servers, Multimedia Telephony Presence Management, and Instant Messaging. M2M Cloud provides device to gateway and Cloud solutions for federation and data and object management. M2M Cloud also provides tools and simulators for sensors for data traffic along with remote SIM card solutions.

Terminals: At the terminals, a plethora of 4G/LTE terminals as well as prototypes, SIM cards to access the 4G/LTE/IMS services, plethora of machine-to-machine sensors, and SIM cards for machine-to-machine tests on real network are provided.

4G/IMS Testbed is a IPV6/IPV4 supported network with Wi-Fi calling and SMS over IMS capability. It also supports tracing and logging. Radio channels are available with frequency division duplex (FDD)/time division duplex (TDD) support. User simulation supported on radio access network (RAN). 4G user equipment like phone, tablets, and USB dongles are used.

Gateways and sensors: Sixty sensors are available to build machine-to-machine test case or proof of concept. Here are a few of the supported sensors. Thermal current flow sensor, sunlight measurement, rain measurement, and wind speed measurement sensors, crack measurement in concrete sensors, concrete temperature measurement sensor, carbon dioxide detection and noise pollution detection sensors. The gateway supported are 2G/3G/4G/Internet.

REFERENCES

1. https://csrc.nist.gov/publications/detail/sp/800-145/final.
2. Nancy Ambritta P., Poonam N. Railkar, and Parikshit N. Mahalle. Proposed Identity and Access Management in Future Internet (IAMFI): A Behavioral Modeling Approach, *Journal of ICT, 2*(1), 1–36, 2014.
3. Lu Hou, Shaohang Zhao, Xiong Xiong, Kan Zheng, Periklis Chatzimisios, M. Shamim Hossain, and Wei Xiang. Internet of Things Cloud: Architecture and Implementation, *IEEE Communications Magazine — Communications Standards Supplement, 54,* 32–39, 2016.

4. Hany F. Atlam, Ahmed Alenezi, Abdulrahman Alharthi, Robert J. Walters, and Gary B. Wills. Integration of Cloud Computing with Internet of Things: Challenges and Open Issues, *IEEE International Conference on iThings-GreenCom-CPSCom-SmartData*, *17*, 670–675, 2017.

5. Ademir F. da Silva, Ricardo L. Ohta, Marcelo N. dos Santos, and Alecio P.D. Binotto. A Cloud-based Architecture for the Internet of Things targeting Industrial Devices Remote Monitoring and Control, IFAC-PapersOnLine, *49*(30), 108–113, 2016.

6. Dimitrios Kelaidonis, Panagiotis Vlacheas, Vera Stavroulaki, Stylianos Georgoulas, Klaus Moessner, Yuichi Hashi, Kazuo Hashimoto, Yutaka Miyake, Keiji Yamada, and Panagiotis Demestichas, *Cloud Internet of Things Framework for Enabling Services in Smart Cities*, In: Angelakis V., Tragos E., Pöhls H., Kapovits A., Bassi A. (eds) Designing, Developing, and Facilitating Smart Cities, pp. 163–191, Springer, Cham, 2017.

7. Yu Liu, Kahin Akram Hassan, Magnus Karlsson, Zhibo Pang, and Shaofang Gong, A Data-Centric Internet of Things Framework Based on Azure Cloud, *IEEE*, *7*, 53839–53858, 2019.

8. Laura Belli, Simone Cirani, Luca Davoli, Lorenzo Melegari, MàriusMónton, and Marco Picone, *An Open-Source Cloud Architecture for Big Stream IoT Applications*, In: Interoperability and Open-Source Solutions for the Internet of Things, pp.73–88, Springer, Cham, 2015.

9. Tian Wang, Guangxue Zhang, Anfeng Liu, Md ZakirulAlam Bhuiyan, and Qun Jin. A Secure IoT Service Architecture with an Efficient Balance Dynamics Based on Cloud and Edge Computing', *IEEE*, *6*, 4831–4843, 2018.

10. Asad Javed, Jerémy Robert, Keijo Heljanko, and Kary Främling. IoTEF: A Federated Edge-Cloud Architecture for Fault-Tolerant IoT Applications, *Journal of Grid Computing*, *18*, 57–80, 2020.

11. Lorenzo Carnevale, Antonio Celesti, Antonino Galletta, Schahram Dustdar, Massimo Villari. From the Cloud to Edge and IoT: a Smart Orchestration Architecture for Enabling Osmotic Computing, *IEEE 18*, 419–424, 2018.

12. Aliza Jamal, Farhan Ahmed Siddiqui, Adnan A. Siddiqui, Nadeem Mahmood, Muhammad Saeed, and Syed Asim Ali. Algorithms and Techniques for Computation Offloading in Edge Enabled Cloud of Things (ECoT) – A Primer, *International Journal of Computer Science and Network Security*, *19*(6), 1–11, 2019.

13. Ankan Ghosh, Osman Khalid, Rao N. B. Rais, Amjad Rehman, Saif U. R. Malik, and Imran A. Khan. Data offloading in IoT Environments: Modeling, Analysis, and Verificatin. EURASIP Journal on Wireless Communications and Networking, *53*, 2019.

14. Azham Hussain, S. V. Manikanthan, T. Padmapriya, and Mahendran Nagalingam, Genetic Algorithm Based Adaptive Offloading for Improving IoT Device Communication Efficiency, Wireless Networks 26, 2329–2338, 2020 .

15. P. Kortoçi, L. Zheng, C. Joe-wong, M. D. Francesco, and M. Chiang. Fog-based Data Offloading in Urban IoT Scenarios, IEEE INFOCOM 2019 – IEEE Conference on Computer Communications, pp. 784–792, 2019.

16. Youngjae Park, Sungwook Kim. Game-Based Data Offloading Scheme FORIOT System Traffic Congestion Problems, *EURASIP Journal on Wireless Communications and Networking, 192*, 2015.

17. Rahbari, D., Nickray, M. Task offloading in mobile fog computing by classification and regression tree. *Peer-to-Peer Networking and Applications 13*, 104–122, 2020.

18. Xiaolong Xu, Qingxiang Liu, Yun Luo, Kai Peng, Xuyun Zhang, Shunmei Meng, and Lianyong Qi. A computation offloading method over big data for IoT-enabled cloud-edge computing, *Future Generation Computer Systems, 95*, 522–533, 2019.

19. Qiao G., Leng S., and Zhang, Y. Online Learning and Optimization for Computation Offloading in D2D Edge Computing and Networks, *Mobile Networks and Applications*, 2019.

20. Kim, S. Nested game-based computation offloading scheme for Mobile Cloud IoT systems. *EURASIP Journal on Wireless Communications and Networking, 229*, 2015.

21. A.T. Saraswathi, Y.R.A. Kalaashri, and S. Padmavathi, *Dynamic Resource Allocation Scheme in Cloud Computing*, In: Procedia Computer Science, vol. 47, pp. 30–36, 2015.

22. Pešić S., Tošić M., Iković O., Ivanović M., Radovanović M., and Bošković D. *Context Aware Resource and Service Provisioning Management in Fog Computing Systems*. In: Ivanović M., Bădică C., Dix J., Jovanović Z., Malgeri M., Savić M. (eds) *Intelligent Distributed Computing XI*. IDC 2017. Studies in Computational Intelligence, vol. 737. Springer, Cham, 2018.

23. Avasalcai C., Dustdar S. *Latency-Aware Distributed Resource Provisioning for Deploying IoT Applications at the Edge of the Network*. In: Arai K., Bhatia R. (eds) Advances in Information and Communication. FICC 2019. Lecture Notes in Networks and Systems, vol. 69. Springer, Cham, 2020.

24. Badawy M.M., Ali Z.H., and Ali H.A. QoS provisioning framework for service-oriented internet of things (IoT). *Cluster Computing 23*, 575–591, 2020.

25. Kirkham T. et al. (2016) *Privacy Aware on-Demand Resource Provisioning for IoT Data Processing*. In: Mandler B. et al. (eds) Internet of Things. IoT Infrastructures. IoT360 2015. Lecture

Notes of the Institute for Computer Sciences, Social Informatics and Telecommunications Engineering, vol. 170. Springer, Cham.

26. Vinueza Naranjo P.G., Baccarelli E., and Scarpiniti M. Design and energy-efficient resource management of virtualized networked Fog architectures for the real-time support of IoT applications, *Journal of Supercomputing 74*, 2470–2507, 2018.

27. Yousefpour A. et al. FOGPLAN: A Lightweight QoS-Aware Dynamic Fog Service Provisioning Framework, *IEEE Internet of Things Journal*, 6(3), 5080–5096, 2019.

28. Zhao S., Yu L., and Cheng B. An Event-Driven Service Provisioning Mechanism for IoT (Internet of Things) System Interaction, *IEEE, 4*, 5038–5051, 2016.

29. Sha K., Errabelly R., Wei W., Yang T.A., and Wang Z. EdgeSec: Design of an Edge Layer Security Service to Enhance IoT Security 2017 IEEE 1st International Conference on Fog and Edge Computing (ICFEC), Madrid, Spain, pp. 81–88, 2017.

30. Hsu R., Lee J., Quek T.Q.S., and Chen J. Reconfigurable Security: Edge-Computing-Based Framework for IoT, *IEEE Network*, 32(5), 92–99, 2018.

31. Mukherjee B., Neupane R.L., and Calyam P. End-to-End IoT Security Middleware for Cloud-Fog Communication, IEEE 4th International Conference on Cyber Security and Cloud Computing (CSCloud), 2017, pp. 151–156.

32. Hongxin Hu, Wonkyu Han, Gail-Joon Ahn, and Ziming Zhao. FLOWGUARD: building robust firewalls for software-defined networks. In Proceedings of the third workshop on hot topics in software defined networking (HotSDN '14). Association for Computing Machinery, New York, NY, USA, 97–102, 2014.

33. T. Markham and C. Payne. Security at the network edge: a distributed firewall architecture, Proceedings DARPA Information Survivability Conference and Exposition II. DISCEX'01, Anaheim, CA, USA, pp. 279–286, vol. 1, 2001.

34. Kewei Sha, T. Andrew Yang, Wei Wei, and Sadegh Davari. A survey of edge computing-based designs for IoT security, *Digital Communications and Networks*, 6(2), 195–202, 2020.

35. FIESTA-IoT, www.fiesta-iot.edu

4

SMART COMPUTING
OVER IOT–CLOUD

4.1 INTRODUCTION

Smart computing in general refers to the mechanism of empowering devices and utilities that we use in our day-to-day lives with computing capabilities. On similar lines, smart computing over the IoT–Cloud refers to confluence of hardware, software, and network technologies that empower the IoT–Cloud application with real-time awareness of the environment and enhanced analytics that can assist humans in better decision making and optimizations, thereby driving the business to success.

This chapter discusses smart computing technologies that are involved in driving the IoT–Cloud applications to success. Section 4.1 provides an overview of big data analytics and cognitive computing from an IoT– Cloud perspective. Section 4.2 provides a brief explanation of a few deep learning approaches. Section 4.3 discusses the techniques, algorithms, and methods and Section 4.4 includes a few case studies pertaining to smart computing over IoT–Cloud.

4.2 BIG DATA ANALYTICS AND COGNITIVE COMPUTING

From the earlier chapters, we now have a clear understanding about IoT–Cloud. With the innumerable devices that are connected to each other and the Cloud through the Internet, the amount of data that these devices collect or generate is immeasurable and huge. The size of this data can be in petabytes of gigabytes. Additionally, given the heterogeneity of these devices, platforms, and environment of operation, the data that these devices collect are present in different formats. They come in a variety of forms such as structured (an organized format), unstructured, and semistructured. This huge volume of data in different formats is referred to as big data and analysis of this data in order to generate inferences and solutions is called as big data analytics. However, analysis of such data in IoT applications is a huge challenge due to the their size and heterogeneity.

Cognitive computing is a mechanism used in solving problems that are complex and may have a certain degree of uncertainty in arriving at suitable answers. It is a self-learning system that mimics the human brain/thinking with the help of computerized models. It is a confluence of several underlying technologies such as natural language processing (NLP), pattern recognition, data mining, sentiment analysis, machine learning, neural networks, and deep learning, which we will discuss in detail in this section. The confluence of big data analytics and cognitive computing will help unleash the greater potential of IoT–Cloud applications. This section will discuss about the cognitive computing capabilities, provide a detailed description of the various technologies used and how cognitive computing empowers analytics.

4.2.1 Cognitive Computing Capabilities

In today's IoT-connected world, the number of data generators (human, machines, and other business applications) are multiplying rapidly, consequently generating huge volumes of structured and unstructured data. This has led to the development of an economy that is data-driven. This has indirectly forced many businesses to venture into cognitive computing and enhanced analytics in order to stay relevant and competent. Cognitive computing capabilities are immense and multidisciplinary.

Cognitive computing can offer *improved data analysis.* For instance, the health care industry assimilates data from various sources such as journals, medical records, diagnostic tools, and other documents. All these data provide evidence and help make informed decisions and recommendation related to the treatment that can be provided to patients. Here is where cognitive computing comes in handy by performing quick and reliable analysis of the data and presenting it to the physicians, surgeons, or medical professionals.

Cognitive computing can lead to *improved customer satisfaction levels.* For instance, the Hilton group [1], which is a hospitality and travel business, has employed a robot, Connie (Watson enabled) that provides customers with precise, relevant, and accurate information on various topics related to travel and accommodation. It also provides information on fine dining, amenities offered at hotels, and places to visit thus making the customers have a smart, easy, and enjoyable travel experience.

Cognitive computing can simplify complex processes into simpler and *efficient processes*. In the case of Swiss Re [1], an insurance company, the application of cognitive computing has made the process of identifying patterns simpler and efficient, thereby enabling real-time problem solving for more efficient responses. It has employed the IBM Watson technology to perform analysis of huge volumes of structured and unstructured data pertaining to the risk of exposure of sensitive information. Based on the analysis, measures were adopted to put efficient risk management tools in place and improve the productivity of the business.

Employing cognitive computing can help improve businesses by dealing with *financial services* for purchases. For instance, customers wanting to buy automobiles can have varying levels of financial packages and services. The agents can employ cognitive technologies with the consent of customers in order to surf through their data such as credit score, and other important financial documents in order to suggest offers that would benefit both the buyer and seller.

Cognitive computing can be employed for *identifying safety concerns in a product earlier* in the lifecycle, thereby helping to *reduce costs* that might be incurred in a recall after completion. It also helps in upholding reputations of big organizations by identifying shortcomings at an earlier stage. Also, the delays in time-to-market that might occur if a product fails are also taken care of with early detection.

The banking sector has reaped great benefits by employing cognitive computing. In particular, *fraud detection* has an improved edge where the earlier rule-based detection mechanism (based on standard questions) has been expanded to mechanisms that analyze banking behavior of customers (spending habits), mechanisms that predict future expenditure or buying based on previous patterns, and many such mechanisms. This helps in detection of any anomaly in the regular patterns, leading to card blockage and alert sent to the legitimate customer.

Cognitive computing over IoT can enable products to *make independent and instantaneous decisions* in businesses without human interference. Fact-based solutions can be provided proactively to drive the entire business process right from engaging in relevant and meaningful conversations with customers to the manufacturing and maintenance of tools and equipment.

Cognitive computing must possess the following features in order to realize the above-mentioned capabilities.

- *Adaptive:* Cognitive computing must be able to keep up with the dynamically changing data, goals and requirements by learning, and updating constantly.
- *Interactive:* Cognitive computing should provide flexibility and ease by allowing users to communicate just the way they would in a real-world human-to-human interaction using voice, gestures, and natural languages.
- *Iterative and stateful:* It should possess the capability of collecting relevant information by asking suitable questions from the user in the event where enough information and requirements are not available in order to describe the problem in question.
- *Contextual:* Cognitive computing should discover and extract relevant information like location, time, and user details pertaining to the problem based on sensory inputs such as gestures, speech, and vision. Cognitive computing should analyze and process real-time and near real-time data.

DeepMind is a sophisticated computer that resembles the short-term memory properties of the human brain. This system has a neural network connected to an external memory. It preservers information/queries as memories and then uses them later to understand and act upon new data that it encounters on its own without any human intervention or monitoring.

The *Zeroth Cognitive Computing Platform* is built upon the visual and audio based cognitive computing capabilities that perfectly resemble the actions and thoughts of a human brain. It enables the system to accurately identify things, recognize and read handwritings, recognize people all pertaining to specific contexts just like a human brain. A few applications of the system could be the automatic adjustment of the camera settings to accommodate to the natural light on a bright sunny or cloudy day, and the adjustment of the microphone setting to cancel out any disturbances or noise in the background to help in improving the sound quality. It can also be applied to detect the levels of stress and other feelings of a person from their voices and send alerts.

Watson is an advanced system with in-built cognitive computing capabilities such as dynamic learning, natural language processing, and hypothesis generation and evaluation. It works on normal queries that are asked by users in simple language and converts them into an appropriate data language used for querying.

The information technology service giants such as TCS, Infosys, Wipro, Cognizant, and HCL Technologies have been successful in *reducing human effort and cutting costs* by developing their own technology platforms that incorporate automation, machine learning, and artificial intelligence capabilities. Many tasks have been automated, thereby generating more revenue without much of human effort requirements. TCS' automation platform is Ignio, HCL's is Dry Ice, Infosys operates Infosys Automation Platforms, Wipro manages Holmes, and Cognizant has the Tizentto platform.

Cognitive computing is capable of *minimizing the amount of traffic from and to the Cloud* in an IoT–Cloud system by *imparting intelligence to the edge devices*. Devices can be equipped with capabilities that can reduce energy consumption and improve performance and privacy.

4.2.2 Underlying Technologies

As mentioned earlier cognitive computing is a confluence of many underlying technologies such as natural language processing (NLP), pattern recognition, data mining, sentiment analysis, machine learning, neural networks, and deep learning. This section will provide a brief description of the technologies.

Natural language processing (NLP) [2] is a field of study that helps in translating and interpreting human language by computers. These computers basically work upon the natural human language by analyzing and understanding the language, thereby being able to extract meaningful information in a smart way. A piece of software written by developers using the underlying NLP algorithms can help understand the human language (speech and text) better and use it for analysis. Some of the applications made possible due to NLP's ability to extract meaning from language based on the analysis of the hierarchical structure of language are grammar correction, speech to text convertor, and automatic language translators. NLP algorithms are *machine learning based algorithms* that enable learning of rules of execution by studying/analyzing a predefined set of examples such as books or a set of sentences, leading to a statistically generated inference. Here, the programmer is relieved from the burden of having to write/code the set of rules for analysis.

NLP has a set of open standard libraries that assist in real-time application development. *Algorithmia* is a model based on machine learning that supports deployment and management of applications without the need to spend efforts in setting up servers and infrastructure. It helps a great deal in automating the machine learning operations for an organization with simple API endpoints to the algorithms, some of which are discussed below.

Apache OpenNLP is a toolkit employed to process text that is written in the natural language and helps in development of services that support proficient text processing actions. Common tasks that are performed by NLP like language recognition, segmentation, parsing, and chunking tokenization are supported by this open-source library that is based on machine learning.

Natural Language Toolkit (NLTK) is a collection of efficient libraries for processing text in natural language (English). It is a platform that supports symbolic and statistical NLP with programs written in Python. NLTK has been found more suitable for teaching and research. It is also suitable for empirical linguistics in Python, machine learning, artificial intelligence, and retrieval of meaningful information.

Stanford NLP is a package of NLP software developed and managed by the Stanford NLP group. The tools offered by the group can be integrated into applications that require human language processing, analysis, and interpretation requirements. Its use has been extensive in the fields of academia and in industrial and governmental organizations.

MALLET is a package that is open source and written in Java. It supports statistical NLP with refined tools for various NLP activities such as sequence tagging, numerical optimization, topic modeling, information extraction, and beyond. One of the most easily understandable and relatable examples of NLP is the analysis of text for the tone of message and then marking them as positive, negative, or neutral as done in social media platforms such as Facebook or Twitter. Tracking of topics that are currently trending and popularity tracking of hashtags, auto filtering of offensive text and comments, and chatbots are all efficient implementations of NLP.

Data mining is the process of excavating huge volumes of data in order to draw inferences, patterns, knowledge, and information that can be used to make improved business decisions, devise effective cost-cutting strategies, and improve revenue. It involves the application of certain mechanisms to help in finding anomalies and correlations in larger data sets, thereby enabling detection of outcomes [3]. It is one of the phases in the process of "knowledge discovery in databases". Data mining has six classes of tasks that are performed during the process as listed below.

- *Anomaly detection* is a mechanism applied in order to examine the data and detect any change or outlier or deviation that can be used for further analysis and investigation.
- *Association rule learning* or also known as market basket analysis is a method of identifying relationships, patterns in the data, and associated variables. For example, identification of customer buying habits can help business in understanding frequently bought items and items bought together. This can help in developing efficient marketing strategies.
- *Clustering* is an action of identifying similarities in data, thus leading to the detection of groups and structures in the data. This grouping is done without any predefined/known labels or structures in the data.
- *Classification* is similar to that of clustering in that both the methods perform grouping of data based on identified features. However, classification is a supervised learning technique wherein the data is categorized based on previously available labels or structures, for example, classification of emails as spam and valid.
- *Regression* is a predictive analysis mechanism wherein modeling of data is done in order to assess the strength and relationship between variables or data sets.
- *Summarization* is a representation of data sets in a compressed and compacted way with the help of reports and visualizations (graphs and charts).

Machine learning is a mechanism that imparts knowledge into machines/computer, thereby enabling them to act faster and effectively. The ability to self-learn and improve its response and functioning based on the experience and with no external inputs (programming) is the crux of machine learning built on the idea of artificial intelligence [4]. There are basically three approaches toward machine learning based on the type of stimuli/signal and feedback that the learning agent works upon.

Supervised learning is a mechanism wherein the learning agent is provided some inputs, rules, and expected outcomes as examples from which it is expected to learn and then apply them in the actual operational environment.

Unsupervised learning is a mechanism wherein the learning agent is not provided with any kind of prior knowledge and the agent is given the liberty to act on the environment and generate outcomes on its own based on its own analysis.

Reinforcement learning [5] involves a feedback/reward mechanism. The agent communicates with the real time dynamically changing environment in order to achieve a goal and at regular intervals feedback/rewards are presented based on the progress, which the agent tries to maximize in its journey toward accomplishing the goal.

Neural networks are systems that take its root from a human brain. They are similar to machine learning models in that they learn from experience without the need for programming that is specific to the task. However, neural networks are different in that they are able to make intelligent decisions on its own, unlike machine learning where decisions are made based on what it has learned only.

Neural networks have several layers that are involved in the refinement of output at each level. The layers have nodes that are responsible for carrying out the computations. It mimics the neuron in the human brain wherein it collects and unites the input from the data and assigns suitable weights which are responsible to either magnify or diminish the input value pertaining to the task. An activation function is assigned to a node that decides about the depth of progress that is to be made by the signal through the various layers in order to affect the outcome based upon the sum of the product of the input–weight pair. A neuron is activated in each layer when a signal reaches or propagates through it.

Sentiment analysis finds its application in many well-known platforms and applications. It is basically a text analysis mechanism that is used to detect the polarity (positive, negative, or neutral opinion) in a given text, be it a long document, paragraph, or a few sentences [6]. It basically helps in understanding human emotions from the text. There are various types of sentiment analysis, and they are described below.

- *Fine-grained sentiment analysis* is used when the precision of sentiment/polarity is very important for the business. For example, in a five-star rating for a product or a review for a session, the polarity is recorded in five degrees (Excellent – 5, good – 4, fair – 3, meets expectations – 2, below expectations – 1).

- *Emotion detection* is a method applied to identify human emotions such as happy, sad, angry, or frustrated from a text using a variety of machine learning algorithms.
- *Aspect-based sentiment analysis* is a mechanism that helps identify emotions expressed pertaining to specific features or characteristics of a product or artifact. For example, when customers review a phone, they may specify certain features of the phone as being outstanding or abysmal.
- *Multilingual sentiment analysis* is a method applied on texts written in a variety of languages. This involves a lot of effort in terms of pre-processing and resource requirements. Many tools are also available online that can be used effectively, but for more accurate and specific results, algorithms and techniques must be developed indigenously.

Patten recognition is a mechanism applied to analyze and process data in order to generate meaningful patterns or regularities. Pattern recognition has its application in various areas such as image analysis, bioinformatics, computer graphics, signal processing, image processing, and many more [7]. Pattern recognition techniques are of three types as follows.

In *statistical pattern recognition*, patterns are described using features and measurements. Each feature is represented as a point in an n-dimensional space. Statistical pattern recognition then picks features and measurements that permit the pattern vectors to fit in various groups or categories in the n-dimensional space.

Syntactic pattern recognition uses basic subpatterns or simply referred to as primitives that are utilized in making descriptions of the patterns by organizing them into words and sentences.

Neural pattern recognition works on a system containing numerous processors that are interconnected and enable parallel computing. The systems are then trained on sample sets, enabling them to learn from the given input–output data sets, thereby enabling them to adapt themselves to the changing data.

Deep learning enables machines to "learn by example." A machine acquires the ability to carry out classification by learning from images, sounds, or text messages directly. Deep learning provides unprecedented levels of accuracy that even exceeds normal human ability. Data that are labeled are used to train the models using the several layers in the neural networks. Deep learning will be discussed in detail in Section 4.3.

4.2.3 Empowering Analytics

In the introduction of this chapter, we had presented an apt definition for big data and cognitive computing. The huge volumes of the big data pose huge problems too. Industries and businesses today are overwhelmed with the amount of information that is accumulated for analysis. However, the talent that is required to handle such data and retrieve meaningful information for businesses is scarcely available. The number of data scientists and analysts are not enough to keep up with the ever-increasing data volumes. Experienced specialists are required in order to handle the available platforms and put them to effective use.

A solution to this problem would be to increase the supply of specialists by increasing the training programs offered to interested people. On these grounds, an even more effective solution would be to utilize the existing technologies and train machines/computers rather than human beings to manage the tools. This is made possible with advancements in cognitive computing. The confluence of cognitive computing, artificial intelligence, and machine learning can aid both experienced and inexperienced staff to handle complex analytic operations using the available tools and platforms. It also helps in improving the accuracy and quality of the results. This accelerates the process of analysis in real-time and near real-time data, thereby enabling businesses to make real-time decisions.

The capabilities that cognitive computing can empower big data with are countless and very promising. The argument presented above about the lack of sufficient talent with knowledge to handle the big data platforms can be overcome with advancements in NLP. With NLP in picture, employees who are not proficient in data/information processing and data languages that is required for analytics activities can simply work on the platforms and tools with normal interactions just as we do with other human beings. The platforms can be equipped with the capability to transform normal language into data queries and requests and respond with solutions or answers in the same way as a natural language, enabling easy understanding. This brings in much more *flexibility* into big data analytics.

Big data analytics empowered by cognitive computing has accelerated the decision-making process, accuracy, and productivity of many businesses with its tools and platforms. A few of them are DeepMind, Watson, and Zeroth Cognitive Computing Platform, which we have already discussed in Section 4.1.1. Following are a few more tools that drive the analytics process in many businesses.

IBM Watson has partnered with an Indian health care organization's oncology department based in Manipal. It aids in making better decisions in cancer treatment based on the previous knowledge that it has acquired by going through a vast number of medical records, patient history, and journals to arrive at an accurate diagnosis and recommended treatments.

Wipro's Holmes Platform mentioned in Section 4.1.1 is an artificial intelligence based platform. It has been adopted in the financial sector in order to study, examine, and analyze a variety of structured and unstructured data and support the decision-making process of an analyst. Tasks such as know your customer (KYC) and appraisal pertaining to credit risk are automated with the help of Wipro Holmes.

The government of Andhra Pradesh has sought the predictive services offered by Azure in order to identify students with a high probability of dropout from schools. The platform enables the collection and analysis of data for specific information such as school infrastructure, teacher quality, training methodology, and also the social and economic background. This enables the government in making better decisions and allocate resources where in need.

4.3 DEEP LEARNING APPROACHES

Deep learning is a subset of machine learning in which the age-old traditional algorithms used to instruct the computers on the tasks to be performed are equipped with capabilities to modify their own instructions to improve the functionality of the computer. This is the crux of artificial intelligence. While machines have great advantages of accuracy and speed, they are limited by the program parameters. In order to make the machines more competitive, they should be made more adaptable like the human brain. Deep learning is a mechanism that enables computers to process huge volumes of data and learn from experience similar to humans. The algorithms and mechanisms in deep learning perform tasks in a repetitive manner, each time adjusting the parameters in order to achieve the desired outcome. This section discusses a few approaches in deep learning.

4.3.1 Artificial Neural Networks (ANN)

ANN is a deep learning approach that imparts artificial intelligence into machines by mimicking the human brain and the nervous system. Imagine that you have just hurt your index finger. The sensory nerve in the hand immediately sends out signals (chain reaction) that ultimately reaches your brain and tells that you are experiencing pain. This is the basic idea behind the functionality of ANNs. The ANN have a sequence of branching nodes which function in a similar way to that of the neurons in the human body. Inputs are fed into the input nodes, which then propagate the information into the series of internal nodes, which process the information until the desired output is generated.

Neural networks revolve around the two most vital components of the brain, namely, the neuron and the synapse. The neuron is responsible for processing the signals that it receives from other neurons. The synapse connects the neurons enabling information to be passed on from one neuron to another. In ANN, a node depicts the neuron which receives the information and transforms the information by performing a quantitative function, which is carried over to the next neuron. In transit the connectivity lines (synapse) in turn apply its own transformation function on the information and modifies it by adding a constant value. This modification process happens by the *application of weights*. The input from multiple synapses or connectivity lines are collected, summed up, and then sent to the next node. This node in turn adjusts the data by applying constants, thereby modifying the data. This application of constants at the nodes is called as *node bias*. Application of weights and bias to the input data is important as this ascertains that the data is propagated properly through the network. For a node to be able to propagate/pass the data, it must be activated. Activation of nodes happens when the output it produces meets the threshold value that is set by the programmer, after which the data will be passed on to the next node; otherwise the node remains dormant. However, this single pass of information to the final nodes in many cases might not lead to the desired output. For example, the network might accidentally identify a cat as a dog, which is not acceptable. To counter this, an algorithm called the backpropagation algorithm is applied to the network, which uses feedbacks to enable the adjustment of weights and biases and fine-tune the synapses until the result is agreeable or even almost correct. The backpropagation algorithm is discussed in detail in Section 4.3.

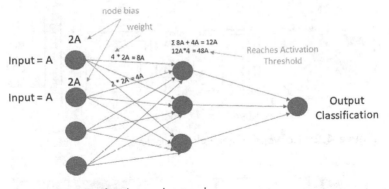

Figure 4.1 Artificial neural network.

Figure 4.1 shows a representation of the artificial neural network as described above.

4.3.2 Convolution Neural Network (CNN)

CNN is mostly applied to image processing problems and natural language processing problems. Traditional neural networks make no assumptions on the inputs and weights that are used to train the models which are not suitable for images and natural language-based problems. CNN treats data as being spatial. As opposed to neurons being connected to neurons in the preceding layer, they are instead linked to neurons that are only close to it and all of them have the same weights. This simplification enables to maintain the spatial property of the data set. The simplification of an image is made possible via the CNN facilitating better processing and understanding of the images [8].

The CNN architecture as shown in Figure 4.2 consists of multiple layers that exist in a usual neural network. The CNN has additional layers such as the convolution layer, pooling layer, ReLu (rectified linear unit) layer, and a wholly connected layer. The ReLu layer functions as an activation layer that ensures nonlinearity while the data moves through each layer of the network. Absence of this layer could cause the loss of dimensionality that is required to be maintained with the data fed into each layer. The *fully connected layer* performs classification on the data set. The convolution layer performs the most important function in the network. It places a filter over an array of

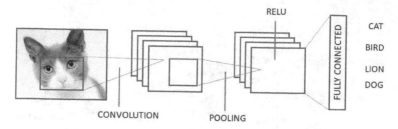

Figure 4.2 Convolution network architecture.

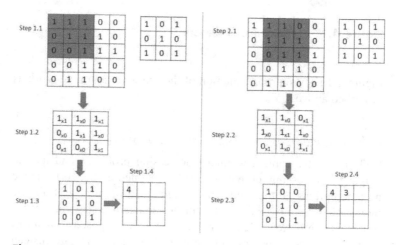

Figure 4.3 Convolved feature map processing.

image pixels, leading to the formation of a "convolved feature map" as shown in Figure 4.3.

This enables focus on specific features of the image, which might be missed out otherwise. The *pooling layer* reduces the number of samples of a feature map, causing a reduction in the number of parameters to process, thereby enabling faster processing. This leads to a pooled feature map. The pooled feature map is obtained by either performing a max pooling (selects the maximum input of a particular convolved feature) or average pooling (calculates the average). Ultimately, the model builds up an image of its own based on its own mathematical rules. If unlabeled data are used, then unsupervised learning methods can be used to *train the CNN*. Auto encoders enable us to compress and place the data into a low-dimensional space.

Calculations are performed in the earlier layers of the CNN followed by reconstruction with the other additional layers, leading to an up sampling of all the data points at hand.

4.3.3 Recurrent Neural Networks (RNN)

RNN are used in many applications such as speech recognition, language translation, and prediction of stocks. RNNs are used to model sequential data. In order to understand this, consider a still snapshot taken of a ball that is in motion over a period [9]. Now, from this picture we would want to predict the direction of motion that the ball is moving in. Such a prediction with a single standstill picture of the ball will be very difficult. Knowledge of where the ball had been before the current picture is required for an accurate and real-time prediction. If we have recorded many snapshots of the ball's position in succession (sequential information over time), we will have enough information to make better predictions. Similarly, audio is sequential information that can be broken up into smaller pieces of information and fed into the RNN for processing. Textual data is another type of sequential information that can be a sequence of alphabets or words. RNNs work upon these kinds of sequential information to provide predictions based upon a concept called "sequential memory." Sequential memory is a mechanism that enables the human brain to recognize patterns that occur in a sequence. This is implemented in RNNs with the help of a looping mechanism that enables to pass on earlier information in the forward direction for processing. This intermediary information is represented as the hidden state, which depicts previous inputs that affect the later states. An overview of a use case can help better understand this looping mechanism in RNNs. Consider building a chatbot that has a goal of classifying users' intentions from the text input provided. This can be handled by encoding the sequence of texts using an RNN. The output received from the RNN is then fed into a standard feed-forward neural network, which will then perform the classification of the intentions. So, consider the following example scenario where user inputs "What is the time?" Figure 4.4 represents the steps 1–5 as mentioned in the algorithm.

INPUT: WHAT TIME IS IT?

1. Break the sentence into individual words as WHAT TIME IS IT?

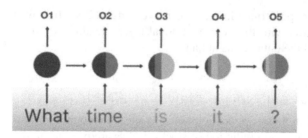

Figure 4.4 Working of a recurrent neural network.

2. Feed the words into the RNN in a sequential manner. Start by feeding in the first word "WHAT" into the RNN. The RNN encodes "WHAT" and produces and output.
3. Next we feed the word "TIME" with a hidden state from the previous step. The hidden state represents information from all the previous steps. Now, the RNN has information on the words "WHAT" and "TIME."
4. Repeat the process for all words until the final step that feeds "?" with all previous state information and produces the final output.
5. Now, since the final stage's (stage 5) output consists of information from all other stages, the output of stage 5 can be taken and provided as input to the feed-forward network that performs a classification of intent on the input.

The RNN, however, suffers from a problem, which is short-term memory and vanishing gradient, which is the side effect of the back-propagation methodology in RNNs. In simple terms, the final output that occurs after sequentially passing in all the input chunks barely carries the memory/information of the output from stage 1, thereby affecting the final output (e.g., the words "WHAT" and "TIME" may not even be considered in predicting the users' intent (short-term memory loss)). Output based on merely "IS" and "IT" may be very ambiguous and unclear. This diminishing information can be handled with advanced RNNs, namely long short-term memory (LSTM) and gated recurrent unit (GRU).

RNN differs from CNN in that, CNN is a feed-forward network used to filter spatial data while the recurrent neural network (RNN) feeds data back into itself, thereby being the best candidates for sequential data. A CNN can recognize patterns across space whereas RNN can recognize patterns over time.

4.4 ALGORITHMS, METHODS, AND TECHNIQUES

Section 4.3 has provided us a brief understanding about deep learning and its approaches. This section will discuss about a few deep learning methods and AI algorithms that are used widely. Following are a few deep learning methods [10] that are commonly used in applications.

Back propagation is a method of training the neural networks that works on the principle of the supervised learning methodology. As explained in Section 4.2.2, neural networks have several layers that process the input to achieve the desired output. While training the model, we provide arbitrary weights at the input that depicts the relevance of each node. The desired output is also specified; however, as these weights are randomly assigned, the weights add up to some amount of error during the process and causes a great difference at the output layer. Backpropagation helps in fine-tuning the weights by reassigning it to an approximate value based on the difference inferred between the actual and desired output. The iterations are repeated until a suitable weight is achieved for the model with minimal error value. The pseudo code is provided below.

1. Values (x,y) are supplied to the input layer nodes.
2. Inputs are assigned arbitrary weights ($w1$, $w2$).
3. The values propagate forward through the model passing all the way through the hidden layer nodes to the output nodes where the output is calculated for every hidden node and the output nodes.
4. The error value (difference between the actual output and the required output) is calculated.
5. Backpropagation from the output nodes to the input nodes is done and, at each node, the weights are adjusted and reassigned so that the error value in step 4 is reduced.
6. The process is repeated for several iterations until the actual output and desired output have minimal or no error (difference in value).

Figure 4.5 shows the working of a back-propagation network. The numbers in the figure represent the steps specified in the pseudo code.

The process of creating a simple neural network and training it using the back propagation algorithm is presented in reference [11].

Stochastic gradient descent is a method that helps in sampling huge volumes of data by randomly selecting data points and sampling them, thereby reducing the amount of computation required. It is

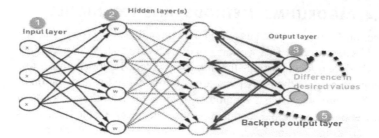

Figure 4.5 Working of a backpropagation network.

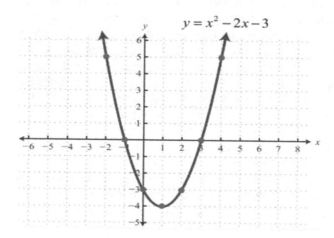

Figure 4.6 Stochastic gradient descent.

based on the unsupervised learning mechanism. A simplified explanation to this method can be given with reference to Figure 4.6. The objective is to find a point x such that the objective function y amounts to zero or minimum. The pseudo code is given below. In the parabola, the lowest point is obtained at $x=1$. However, the objective is to find a value x such that y is minimum ($y=-3$ where $x=0$ or 2).

1. Compute the slope for y (objective function also referred to as the gradient descent function at iterations) with respect to the parameters/features.
2. Assign initial value for the parameters at random (differentiate y with respect to x).
3. Revise the gradient descent function with these new parameter values.

4. Calculate the step size for every feature as step size = gradient * learning rate. Learning rate should be kept to a minimum so that on its journey down the slope we do not want to miss out the minimum point by taking larger strides.
5. Determine new parameter values as new parameter value = old value − step size.
6. Continue the process by repeating steps 3–5 until the gradient reaches a value 0 or close.

Transfer learning [12] is a method adopted to train models in layers by adopting the convolution mechanism. The last few layers tend to be more specific to the data fed as input while the starting layers are more generic pertaining to simple patterns. For example, in a model provided with a training dataset, the early layers might be looking for eyes, ears, and mouth while the later layers may be looking for dogs, humans, and cars.

Following are a few AI algorithms that are most used in applications.

Logistic regression is a classification algorithm working on the principle of supervised learning. It forecasts the probability of a dependent or target variable. The dependent variable has only two outcomes coded as 1 (success/yes) and 0 (failure/no). It is the simplest mechanism used in classification problems such as illness prediction (cancer and diabetes) and spam email identification.

Naive Bayes algorithm is another classification method based on the supervised learning mechanism. The central idea of the naïve Bayes classifier is the Bayes theorem. The classification has two phases, namely the learning phase (model trained on a given dataset) and evaluation phase (performance testing). Given the example of weather conditions and playing a sport, the goal is to determine the probability of playing the sport and also determine if the players will play the sport or not (yes or no) with respect to the weather. The Bayes algorithm is given as follows:

where

- $P(H)$ is the prior probability of H (yes/no), that is, probability of hypothesis being true.
- $P(D)$ is prior probability of D (for example, D can be the weather conditions such as sunny or rainy).
- $P(H|D)$ is probability of H given D and is called the posterior probability.
- $P(D|H)$ is probability of D given H and is also called the posterior probability.

Problem: The players play if the weather is rainy. Determine if the statement is true or false. Refer Table 4.1 for values. Table 4.2 shows the frequencies of the classes and labels. Naïve Bayes classifier determines the probability of an event as follows.

1. Determine the prior probability for the known class labels (for example, yes/no) and for each attribute (D, i.e., (Rainy, Sunny)) as shown in Table 4.3.
   $P(yes) = 2/5 = 0.4$, $P(no) = 3/5 = 0.6$, $P(Sunny) = 3/5 = 0.6$, $P(Rainy) = 2/5 = 0.4$.

2. Find the posterior probability for each attribute (D, i.e., (Rainy, Sunny)) and for every class label (yes,no) as shown in Table 4.4.
   $P(Sunny|No) = 2/3 = 0.67$, $P(Rainy|No) = 1/3 = 0.33$, $P(Sunny|yes) = 1/2 = 0.5$, $P(Rainy|Yes) = 1/2 = 0.5$

3. Substitute the values in Bayes formula and compute posterior probabilities. Probability that players won't play when the weather is rainy is given by
   $P(No|Rainy) = [P(Rainy|No)*P(No)]/P(Rainy) = (0.33*0.6)/0.4 = 0.495$
   and the probability that the players will play given that the weather is rainy is given by
   $P(Yes|Rainy) = [P(Rainy|Yes)*P(Yes)]/P(Rainy) = (0.5*0.4)/0.4 = 0.5$

4. Evaluate the class with higher probability, such that the input belongs to the class with higher probability. In the example, the probability that the players will play when the weather is rainy has a slightly higher probability in comparison to that of the probability the players will not play if it rains. Hence, we can conclude that the hypothesis/problem statement is true.

A similar example [13] as provided above can be referred as a guidance to the implementation of the naïve Bayes classifier. The author has used a simple data set with three columns, namely, weather, temperature, and play for effective demonstration.

Support vector machines (SVM) is a machine learning algorithm that follows the supervised learning mechanism. It is widely used in classification problems where the goal is to find a hyperplane that best segregates all the data points into two different categories as shown in

TABLE 4.1
Naïve Bayes Data Set

Weather	Class Label
Sunny	No
Sunny	No
Rainy	Yes
Rainy	No
Sunny	Yes

TABLE 4.2
Naïve Bayes Frequency Table

Weather	No	Yes
Sunny	2	1
Rainy	1	1

TABLE 4.3
Naïve Bayes Prior Probability

Weather	Yes	No	
Sunny	2	1	=3/5=0
Rainy	1	1	=2/5=0.4
Total	3/5=0.6	2/5=0.4	

TABLE 4.4
Naïve Bayes Posterior Probability

Weather	Yes	No	Posterior Probability (Yes)	Posterior Probability (No)
Sunny	2	1	2/3=0.67	½=0.5
Rainy	1	1	1/3=0.35	½=0.5
Total	3	2		

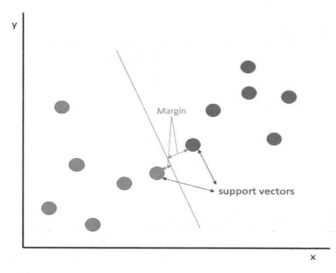

Figure 4.7   Support vector machine.

Figure 4.7. Support vectors are nothing but data points themselves that are closer to the hyperplane whose removal can cause a great change or shift in position of the hyperplane. The success of SVMs is determined from the distance that exists between the hyperplane and the nearest data point, which is called as the margin (greater distance means effective classification). When a clear hyperplane cannot be identified, a 3D view of the data points can help in obtaining the hyperplane, which is done through a mechanism called kernelling. SVM finds its application in many areas such as in the cancer and neurological disease diagnosis and many other research related to health care.

For further understanding, readers can refer to the example in reference [14], which shows the implementation of support vector machine in Python using scikit-learn. The data set used for the implementation is the caner data set, which can be downloaded from the UCI Machine Learning Library. The data set comprising 30 features classifies cancer into two classes (target), namely malignant and benign.

Following are a few machine learning techniques that are used commonly in cognitive computing [15].

Random Forest as the name suggests is a collection of decision trees that when presented with an entity for classification performs an analysis and presents a class to which the entity might belong. Each tree votes for a class independently and a unanimous decision is made

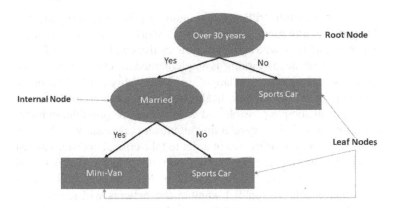

Figure 4.8 CART.

by choosing the class that has the largest number of votes. The decision trees are implemented as a forest in the following steps.

1. If the training data set holds N cases, then a sample space containing N cases are selected randomly with replacement, which will be used to grow the decision tree.
2. For an input with M variables, m variables lesser than M are selected randomly and the best possible split value is used to split the node. This m value remains constant throughout the process of growing the forest.
3. Every decision tree is grown to the maximum size possible without pruning.

CART stands for classification and regression tree. The non-terminal nodes (root node and intermediate nodes) signify a single input variable (x) and a splitting point on the variable. The terminal nodes (leaf nodes) signify the output variable (y). The model predicts the output by walking through the decision tree paths that satisfy the conditions and arrives at the specific output. For example, in Figure 4.8, to know if an unmarried person aged above 30 will buy a minivan or a sports car will follow the path: Over 30 yrs (yes) → Married (No) → Sports car.

4.5 CASE STUDIES

4.5.1 Health Care

As the average life expectancy of humans increases, health care is evolving into multiple specialties and unlimited facets. In fact, health care is a multi-trillion-dollar industry with technology that has changed the way medical professionals work by enabling accurate diagnosis and improved care. It also saves lives and lessens suffering. In health care, deep learning is implemented in every stage of the workflow to elevate the standard of care and quality of life. One of the most important challenges in health care is to get the right treatment to patients as soon as possible and as efficiently as possible. In intensive care units (ICUs), unrelenting attention is necessary with large, incessant data to be processed in order to take critical decisions as fast as possible. Researchers in MIT's Computer Science and Artificial Intelligence lab have developed a deep learning tool called ICU Intervene, which will predict whether the patients will require certain ICU treatments. This model utilizes hourly measurements of very important signs like blood pressure, heart rate, and glucose levels along with patient data like age and gender in order to predict necessary treatments.

Another deep learning model by a startup Corti analyzes phone calls from emergency services. The audio recorded is analyzed and helps emergency personnel to identify if a caller is about to undergo cardiac arrests within a minute. This is also a time-sensitive decision-making process. LexiconAI is a speech recognizing mobile app that can capture medical information with doctor–patient interactions. This enables automatic filling of electronic medical records, thereby enabling doctors to interact more with patients. Millions of medical scans and tissue biopsies are going digital every year. Proscia, a startup uses deep learning with 99 percent accuracy to classify skin pathologies. SigTuple, a startup, has developed an AI microscope that can analyze blood and other fluids of the body. It can analyze the digital images either in the Cloud or at the microscope itself, when compared to digital scanners. SigTuple does it in a fraction of a cost.

To help people with impaired vision, Aira a startup uses AI-guided service used on a smart glass, which helps people navigate and perform their daily routines easily. Aira connects with multiple service agents for assistance. Agents can see what the blind person sees, enabling assistance in just about anything. Aira has a forward-facing camera along with the ability to connect to a smartphone to tap into the GPS and connectivity. To help customers with more complex tasks, human agents are accessible at the tap of a button on the glasses or the app. The agents communicate directly with Aira users via phone, providing real-time visual information with audio support. Complex image recognition and natural language processing tasks are sent to the Cloud for inference. Georgia Tech has introduced an AI prosthetic hand that assisted jazz musician Jason Barnes play piano for the first time in five years. The prosthesis uses electromyogram sensors to recognize muscle movement and allows for individual finger control.

4.5.2 Smart Home

Like in health care, cognitive computing plays a key role in smart homes. Applying machine learning enables smart homes to scale effectively and efficiently. Predictive maintenance is the first key benefit of Cognitive Smart Homes. It can provide critical information to the building owners about equipment failure well ahead in time. Nest Labs' thermostat can also help lessen energy expenses by using sensors that identify the number of occupants in a room by automatically adjusting the temperature. Philips smart light bulbs can turn on and off by detecting occupants in the room. Smart building is connected to smart grid/city and they can listen or talk to each other, thereby making it self-aware. Voice-activated systems contain virtual assistants that can learn and personalize the smart home to their occupants. Smart TVs access content through internet along with voice and gesture support. Yale Smart locks and garage door openers can grant or deny access to visitors. It can automatically unlock the door for its occupants. Smart security cameras enable residents to observe their homes while they are away. Smart sensors can recognize the difference between residents, pets, visitors, and burglars and alert the authorities. Automation of pet care can be done with connected feeders. Plants at home and gardens can be watered by attached timers. Monitors of household systems can detect an electric surge, failures in water system, or freezing pipes and immediately turn off appliances so that the homes are safe. Smart refrigerators monitor the expiry date, make a shopping list, or suggest a recipe based on the contents of the fridge. Smart coffee makers can prepare a fresh cup of coffee. The next logical step of improving the accuracy and capability of IoT sensor driven systems is infusing human awareness via cognitive computing and infuse intelligence into these devices.

4.5.3 Manufacturing

Industrial firms are constantly seeking to bring down operational costs and machine downtime. Cognitive computing offers predictive maintenance solutions that can accurately predict equipment failure. By reducing machine downtime, cognitive computing helps industrial companies to work smarter and safer at the same time, reducing operational expenditure. Predictive maintenance solution from Baker Hughes GE is a deep learning-based model designed to predict equipment failure as early as two months in advance.

In early twentieth century, cars were sold with essentially no options. More than a century later, almost 99 percent of vehicles ordered are unique with customizations. In other words, every order is a custom, so the logistics team in the automobile industry must take in millions of raw parts a day from thousands of suppliers and different locations and organize them into custom-part trays for every customer order. Multiple models of cars can be assembled at the same time and a car is output every minute and the production line can never stop. Achieving this high level of accuracy in factory logistics requires AI-powered industrial logistics. Smart robots in BMW are autonomous in nature, which pick up the car trays and move them to the manufacturing areas. These smart robots perform perception, 3D human pose estimation, localization, actuation, and path planning performing trillions of operations per second of performance.

4.5.4 Retail Sector

Amazon Go is a prime example of the confluence of Cloud and IoT put into action in the retail sector. It is the consumer store of the future where customers can simply walk into the store, take what they want, and come out. After a few minutes, users get a bill for all things they have taken from the store. Amazon Go took three years just to reach the proof-of-concept stage and another three years of rigorous testing and development to reach production stage and opening up the product to customers. It belongs to a class of NP-complete algorithmic problem.

The heart and soul of the Amazon Go Store is computer vision-based machine learning. It is a highly scalable system that can effectively track and identify the objective of all the customers in the store. Amazon Go had to solve six fundamental problems to provide this seamless experience. It is called as the Just Walk Out Technology. They are as follows.

Sensor Fusion represents the problems involved in combining signals coming from different IoT devices in real time. This is very important because if two or more people come close together a problem of tangled state occurs. It means that the AI system has low confidence of who took what and who is who. Another problem is occlusion where people or objects could be hidden from the view.

Calibration represents the problem associated with each IoT device knowing its location precisely in the store. There could be items that are very similar, say, the different flavors of tea from the same brand. Lighting conditions in the store and deformation can change the looks of the item.

Person Detection involves the ability to track and identify individual persons throughout his or her stay in the store. Here it is highly important that the labels are preserved across the video when the customer is moving across the store.

Object Detection involves the ability to identify all the items that are being sold.

Pose Estimation represents the ability to identify each customer's activity near the shelf, especially monitoring the movements of their hands.

Activity Analysis involves checking if the customer has returned/ replaced the item on the shelf or picked it up.

For the computer vision to be successful, it needs Cloud for computation and the magic will not happen until the video feed does not reach the Cloud quickly. Here the IoT devices play a vital role. It does prepossess before streaming it to the Cloud, thereby reducing the video stream size using novel methods such as compression using video codecs. The IoT devices must be resilient enough to have a fail-safe mechanism in case of device failure or network failure such as a backup mechanism to kick in to ensure that the stream reaches the Cloud.

REFERENCES

1. Danish Contractor and Aaditya Telang, *Applications of Cognitive Computing Systems and IBM Watson*, Springer 2017.
2. M. Filgueiras L. Damas N. Moreira, and A. P. Tomás, "Natural Language Processing," EAIA '90, 2nd Advanced School in Artificial Intelligence Guarda, Portugal, October 8–12, 1990.
3. Chun-Wei Tsai, Chin-Feng Lai, Ming-Chao Chiang, and Laurence T. Yang, "Data Mining for Internet of Things: A Survey," *IEEE Communications Surveys & Tutorials*, *16*, 77–97, 2014.

4. Mario Bkassiny, Yang Li, and Sudharman K. Jayaweera, "A Survey on Machine Learning Techniques in Cognitive Radios," *IEEE Communications Surveys & Tutorials*, *15*, 1136–1159, 2013.
5. Akira Notsu, Katsuhiro Honda, Hidetomo Ichihashi, and Yuki Komori, "Simple Reinforcement Learning for Small-Memory Agent," 10th International Conference on Machine Learning and Applications and Workshops, 2011.
6. Neha Raghuvanshi and J. M. Patil, "A Brief Review on Sentiment Analysis," 2016 International Conference on Electrical, Electronics, and Optimization Techniques (ICEEOT), 2016.
7. M. Basu, H. Bunke, and A. Del Bimbo, "Guest Editors' Introduction to the Special Section on Syntactic and Structural Pattern Recognition," IEEE Transactions on Pattern Analysis and Machine Intelligence, 2005.
8. Guangxin Lou and Hongzhen Shi, "Face Image Recognition Based on Convolutional Neural Network," IEEE 2020.
9. Mohammad Reza Mohammadi, Sayed Alireza Sadrossadat, Mir Gholamreza Mortazavi, and Behzad Nouri, "A Brief Review over Neural Network Modeling Techniques," IEEE International Conference on Power, Control, Signals and Instrumentation Engineering (ICPCSI), 2017.
10. S. Durga, Rishabh Nag, and Esther Daniel, "Survey on Machine Learning and Deep Learning Algorithms used in Internet of Things (IoT) Healthcare," 3rd International Conference on Computing Methodologies and Communication (ICCMC), 2019.
11. "Backpropagation in Neural Networks: Process, Example & Code." Neural Network Concepts. https://machinelearningmastery.com/implement-backpropagation-algorithm-scratch-python/.
12. Ling Shao, Fan Zhu, and Xuelong Li, "Transfer Learning for Visual Categorization: A Survey," *IEEE Transactions on Neural Networks and Learning Systems*, *26*, 1019–1034, 2019.
13. www.datacamp.com/community/tutorials/naive-bayes-scikit-learn
14. www.datacamp.com/community/tutorials/svm-classification-scikit-learn-pythony
15. S. Athmaja, M. Hanumanthappa, and Vasantha Kavitha, "A Survey of Machine Learning Algorithms for Big Data Analytics," International Conference on Innovations in Information, Embedded and Communication Systems (ICIIECS), 2017.

5

CONCLUSION

5.1 SUMMARY

This section concludes the book and proposes the future outlook, which can be explored further to initiate new research avenues. This book intended to cover requirements, challenges, technologies, and methods to be adapted in order to bring about the convergence of IoT and Cloud for smart computing. The main topics covered in this book are overview of IoT in terms of architecture and use cases, IoT application development and protocol requirements, opportunities and techniques for IoT–Cloud convergence, and smart computing requirements through learning techniques. The chapter-wise details are summarized below.

The advancements in tiny smart devices, IoT, and Cloud computing have laid the basis for a converged service network called IoT. The seamless communication and contextual services are provided by IoT using smart objects. An overview of IoT covering the technical building blocks and the usage of IoT in business have been presented in Chapter 1. It also presents the IoT layered architecture. The chapter proceeds further by providing an overview of smart computing and discusses a few use cases. The chapter concludes with a discussion on the IoT design issues and challenges.

Due to rapid advancements in e-commerce and m-Commerce as well as availability of Internet at faster and cheaper rate, the use of mobile applications and IoT-enabled applications is increasing at faster rate. IoT solutions have become an integral part of everyday life and IoT application development is becoming a potential business proposition. Smartphones, sensors, cameras, and RFID are the major components in all IoT use cases. Minimum human intervention for the exchange of information and interconnecting devices using wired and wireless communication medium are the key steps in building IoT applications. Chapter 2 begins with a brief description about the main principles to be taken into consideration before developing IoT applications. It then explains the IoT application development phases using a novel C model. Later, the chapter introduces the reader to the various wireless technologies that are suitable for IoT application such as Zigbee, WiFi, and Bluetooth. The chapter also provides a detailed explanation about the protocol stack that is used for IoT application development, which are provided by various working groups. Finally the chapter concludes with a brief explanation of different electronic platforms and their comparison.

In small standalone consumer applications like the smart home and care for disabled and elders, the capabilities required such as processing power, storage space, and computing can be facilitated by the participating devices (mobile phones, Raspberry Pi kits) themselves. The resource requirements for such applications are lower. However, for larger commercial real-time applications such as health care, and transportation, the demand for resources is huge due to the enormous data that is collected and also the complexity of computations that are involved in the system. The demand for resources (storage, servers, and processors) in such huge applications many a times cannot be satisfied by the participating devices alone. This is where Cloud comes in handy by catering to larger business needs. In Chapter 3, we primarily discussed about how Cloud plays a vital role in realizing IoT to its full potential. The opportunities and challenges that are encountered when the IoT–Cloud convergence comes into picture are also discussed in this chapter. It provided a bird's eye view of the architecture for enabling this convergence. It proceeds further by elaborating on data offloading and computing from the IoT perspective. The concept of dynamic resource provisioning for IoT with Cloud is also explained. Finally, the chapter introduced the various test beds and technologies with respect to IoT–Cloud convergence application development.

Smart computing in the context of IoT–Cloud leads to the introduction of advanced computing capabilities including real-time awareness of the surrounding environment and analytics (of the big data detected from various sensors and devices) into the IoT applications by leveraging the capabilities of Cloud. This enables incorporating intelligence into the decision-making process with advanced resources made available through the Cloud. Chapter 4 provided an insight into big data analytics and cognitive computing capabilities and the underlying technologies. The chapter further explained the various deep learning approaches that help realize smart computing in a IoT–Cloud system. A few major case studies under various domains where smart computing over IoT–Cloud is used have been elaborated in the chapter enabling the readers acquire a clear and vivid picture of the usage of smart computing over IoT–Cloud.

The revolution in different technologies like Cloud computing, IoT, cognitive computing, big data analytics and deep learning has led to the emergence of smart application and efficient services with less human intervention. This convergence of different technologies, especially Cloud computing, has accelerated and eventually opened up many personal and business opportunities. This chapter provides a summary of each chapter. The chapter also discusses the issues and challenges involved in developing an IoT–Cloud application. It further provides future outlook of Cloud and IoT in several domains, thereby laying a foundation for further research.

5.2 ISSUES AND CHALLENGES

As stated earlier, the convergence of IoT and Cloud opens up numerous opportunities, It also raises new issues and challenges, which arise during the development of IoT applications or connecting these applications to Cloud. IoT devices suffer from a scarcity of resources, which includes limited memory, battery power, and computational power. These devices are equipped with wireless transmitters to initiate communication and a primitive embedded operating system is installed on these devices. The following is a summary of issues and challenges [1,2]:

(a) *Low-speed networks*: For connecting IoT devices to the gateway, there is low-speed network connectivity for data transfer. However, data transfer is huge and there is a need of lightweight data transfer algorithms for efficient data transfer.

(b) *Scalable data handling algorithms*: Data generated by heterogeneous IoT devices are finally aggregated at a central data storage platform and scalable data handling algorithms are required at the server side. The required operations to be performed on the data are sorting, analyzing, and performing machine learning related tasks like prediction or recommendation. These algorithms at the server side need to be scalable in order to support huge numbers of devices connected to IoT applications, which are located at distributed geographical locations.

(c) *Information fusion*: In emerging IoT applications like health care and agriculture, the data comes from multiple types of sensors and there is a need of efficient information fusion techniques for multi-sensory data. In applications like remote sensing and geographic information system (GIS), the relevance of multi-sensor data poses a big challenge and therefore reliable information fusion techniques are needed.

(d) *Security*: Due to distributed and decentralized nature of IoT applications, they are more prone to attacks at various layers. Possible places of attacks are at devices, communication link, and at the Cloud storage, which include attacks on integrity, confidentiality, privacy, and identity. To resist these various types of attacks, we need lightweight security algorithm not for detecting the attacks but for their prevention. Security by design approach is recommended to address this challenge.

(e) *Availability or downtime*: Critical IoT-enabled applications like health care or defense do not afford downtime. A single point of failure situation needs to be avoided during the convergence of IoT and Cloud. Dynamic load balancing solution should be in place for adaptive load transfer from failure node to active node in order to avoid downtime.

(f) *Quality of service*: Quality of service parameters for IoT applications includes latency, packet delivery ratio, and energy consumption and their requirement varies from one use case to another. The quality of service requirement and feasibility analysis should be carried out in the design part of application development to meet to the customers' requirements.

(g) *Protocol development*: Protocol development in IoT suffers from lack of standardization. So the selection of appropriate protocol is very important from the perspective of both developers and customers. There should be also directed efforts in order to standardize IoT protocols so that designers and developers will have freedom to select from a pool of protocols.

5.3 FUTURE OUTLOOK

The focus of this book is to put forth advances in the field of IoT and Cloud convergence. There are still many areas that need further attention and have prospective applications in future. Convergence of IoT and mobile Cloud for health care applications like remote telemetry is an important area which can be researched further.

Next generation mobile and wireless technologies will have greater impact on the technologies and protocols used for the convergence of IoT and Cloud [3]. Utmost care has to be taken to ensure interoperability and compatibility in both upward and backward directions. Due to scalable and efficient features of Cloud, it has become a potential candidate in the application of IoT devices and storage and processing of big data generated from IoT devices. However, data privacy is at risk as these IoT devices are connected through several communication technologies that are both wired and wireless. In addition to this, the corresponding operating system on the smartphone is also prone to data privacy leaks. Therefore, there is a need for new frameworks for privacy-aware data aggregation to ensure security and privacy of next generation smartphone technologies used in integration of IoT to Cloud.

When IoT–Cloud convergence is used for use cases like remote patient monitoring, remote telemetry, or monitoring old age people remotely, then intelligent computing at Cloud side requires more advanced techniques, which include cognitive computing for data science and quantum computing for personalization, which need further research. Furthermore, the potential of quantum computing needs to be extended for big data analytics for providing secure platform for communication and cyber security and enhancing computation power. Weather forecasting is another interesting use case, which can enhance food quality, production, and retail sales. Quantum computing can contribute to this field to a great extent by managing complex arithmetic at a faster rate and in building climate models to predict global warming and greenhouse gases. This can help the society in preventing global environmental disasters by initiating preventive measures. Further, there is also a need of generic framework that will integrate or converge multiple combinations like IoT and Cloud, IoT and Big Data on Cloud, or any other suitable combination of these three technologies.

In addition to this, there are other allied areas like experimenting and validating data science techniques on real-time data posted on the Cloud, mapping various machine learning techniques with respect to the nature and type of data set, and proposing appropriate tools for Cloud data analytics.

REFERENCES

1. P. N. Mahalle and P. N. Railkar. *Identity Management for Internet of Things*. River Publishers, Wharton, TX, 2015.
2. P. Mahalle, "Identity Management Framework for Internet of Things." PhD Dissertation. Aalborg University, Denmark, 2013.
3. M. S. Hossain, C. Xu, Y. Li, A.-S. K. Pathan, J. Bilbao, W. Zeng, and A. El Saddik, "Impact of Next-Generation Mobile Technologies on IoT-Cloud Convergence," *Communications Magazine*, 55(1), 18–19, 2017.

INDEX